ESSENTIALS OF DRAFTING

A TEXTBOOK ON MECHANICAL DRAWING AND MACHINE DRAWING

WITH

CHAPTERS AND PROBLEMS ON MATERIALS
STRESSES, MACHINE CONSTRUCTION AND
WEIGHT ESTIMATING

BY

CARL L. SVENSEN, B.S.

ASSISTANT PROFESSOR OF ENGINEERING DRAWING IN THE OHIO STATE
UNIVERSITY, JR. MEM. A.S.M.E., MEM. S.P.E.E., FORMERLY INSTRUCTOR
IN MECHANICAL ENGINEERING IN TUFTS COLLEGE AND HEAD
OF THE DEPARTMEMT OF MACHINE CONSTRUCTION
AT THE FRANKLIN UNION

SECOND PRINTING—CORRECTED

NEW YORK
D. VAN NOSTRAND COMPANY
25 PARK PLACE
1919

THE·PLIMPTON·PRESS
NORWOOD·MASS·U·S·A

DEDICATED TO THE AUTHOR'S FRIEND
GARDNER CHACE ANTHONY
WHOSE INFLUENCE AS AN ENGINEER
AND·TEACHER ON THE DEVELOPMENT
OF AMERICAN MECHANICAL DRAWING
IS UNIVERSALLY RECOGNIZED

PREFACE

THE evening technical school has been rapidly developing during recent years. From a makeshift it is coming to occupy a field distinctly its own. The ambitious man attending an evening technical school is fully the equal of his brother at the day technical school and his worth is being increasingly realized.

The foundation subjects — mathematics, mechanics, and drawing — require particular attention in evening courses, where the time may be somewhat limited and the needs of the student varied. This book has been prepared for Ohio Technical Drawing School students as part of a technical course.

Progress in engineering work of any kind depends upon an intimate knowledge of mechanical drafting as the language of the engineering world. Its possibilities must be understood. The mere drawing of lines and more or less copying of exercises or sketching from a few models is far from the purpose of a drawing course. The value of drawing as one of the working tools to be treasured and used during a lifetime in the most useful of professions, ENGINEERING, should be realized. It is as an aid in the study, and later use of engineering knowledge, that drawing finds its place. These preliminary remarks may serve to explain the makeup of the book.

The actual handling of the instruments can best be taught by careful individual instruction of each student, after which false or awkward motions should be immediately corrected. Inefficiency in this respect is one of the most severe handicaps of many "self-made" draftsmen. The treatment of the various subjects is necessarily somewhat brief, as it is intended that personal instruction should be given in each subject.

In the first studies the student is taught to represent each object in strict conformity to the laws of projection. All lines are drawn, all intersections are shown, and invisible surfaces are all indicated by dotted lines. For simple parts such drawings are easily read and they are generally used in the drafting room. When more complicated pieces are met with or where whole machines or constructions are to be represented, such a method would lead to great confusion and often would produce a drawing which it would be almost impossible to read. The time necessary would be very great even for an expert. In such cases the full lines representing the visible surfaces are shown, but the intersections and invisible surfaces are not all drawn in. The selection of what lines to draw and what lines to leave out is an important study in itself.

Furthermore there are many representations of parts which are or appear to be violations of orthographic projection, which are used because practice has shown that they convey the idea to the workman more completely or easily. Other representations are used to save the draftsman's time, or in the interests of simplicity. Almost anything which will make a drawing more readily intelligible is justified. This statement must be used with caution, as what will seem plain to a man familiar with the work may not be so plain to the workman or other reader.

A drawing has one great purpose, and that is to be useful. To this end lines may be added or left out, shading may be used, or notes may be put on. As an expression in the engineering language each drawing should have only one meaning, and should state that meaning with the least possible chance for misinterpretation. Many of these idiomatic expressions of the engineering language will be considered in the later chapters.

The chapters on Materials and Stresses, Machine Construction, and Estimation of Weights are brief treatments of subjects which are necessary for the making of intelligent drawings. Considerable elementary machine design is included as belonging in a practical treatment of mechanical drafting, for the author does not look with favor upon fine distinctions between "subjects." It is the "usability" which really counts.

The subjects are arranged to suit the author's convenience, but they may be taken in a different order if desired. The problems are placed in one chapter at the end of the book, so that a selection may be easily made. These problems are suggestive, and may be amplified by the teacher, who should make a free use of actual shop blueprints and castings.

The author believes that the highest grade work can be done by evening school men, and in an experience of many years has always found that they are ever ready to meet the most exacting requirements when satisfied that what they are receiving is really worth while.

Appreciation of the helpful criticisms of Prof. Thos. E. French is here expressed.

CARL L. SVENSEN

Columbus, Ohio,
   Sept. 1, 1917.

CONTENTS

xi

CONTENTS

CONTENTS

ESSENTIALS OF DRAFTING

CHAPTER I

DRAWING INSTRUMENTS AND MATERIALS

Instruments and Materials. — Drawing instruments and materials should be selected with care, and under the guidance of an experienced draftsman or teacher. The really necessary equipment consists of the following:

Set of case instruments comprising:
    6-inch compasses with fixed needle point leg,
    removable pencil leg and removable pen leg,
    5-inch dividers,
    5-inch ruling pen,
    Bow pencil, bow dividers, bow pen.
24-inch T square.
16 ″ × 20 ″ drawing board.
6-inch 45° triangle.
10-inch 30° × 60° triangle.
Irregular curve.
12-inch architect's scale.
One dozen thumb tacks.
2 H and 4 H or 6 H drawing pencils.
Drawing paper.
Erasive and cleaning rubbers.
Pencil pointer.
Black waterproof drawing ink.
Lettering pens and pen holder.
Pen wiper.

Use of the T Square and Triangle. — The T square is used for drawing horizontal lines, with the head always against the left-hand edge of the board, Fig. 1. The upper edge of the T square blade is always used, and lines are drawn from left to right. The

1

triangles are used for drawing all other lines. Vertical lines are drawn by placing a triangle against the upper edge of the T square and drawing upward along the vertical edge, which should be placed toward the head of the T square, as shown in Fig. 2.

Use of the Scale. — The scale is used for laying off distances. Whenever practicable, drawings should be made full size. If a

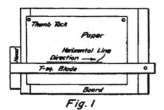

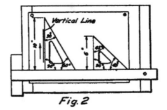

Fig. 1   Fig. 2

reduced scale must be used to accommodate the size of paper, choose one which will show the object clearly, and which will not require great crowding of dimensions. For mechanical drafting, the architect's open divided scale, shown in Fig. 3, is most used. There are many forms, both flat and triangular in section. The following divisions are in general use, $1/8$, $1/4$, $3/8$, $1/2$, $3/4$, 1, $1\frac{1}{2}$, and 3 inches to the foot. The scale $3'' = 1'$ means that the drawing is one fourth the size of the object, or that each

Fig. 3.

one fourth inch on the drawing represents 1 inch on the object. In this case, the 3 inches is divided up into 12 parts, each of which represents 1 inch. These parts are further divided to represent quarter inches and other fractions. The double mark (″) following a figure means inch or inches; the single mark (′) means foot or feet. A common scale graduated to $1/32$ of an inch may be used for many reductions. In such cases use the half inch for an inch in drawing one half size; the quarter inch for an inch in drawing one fourth size, etc. For half size one sixteenth becomes one eighth, and similarly for other divisions.

Fig. 3 shows the distance 2 feet, $5^1/_2$ inches, laid off with the scale of $3'' = 1'$.

Drawing Pencils. — It is necessary to have pencils of the right degree of hardness and properly sharpened. For lettering,

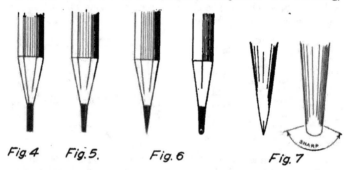

Fig. 4   Fig. 5.   Fig. 6   Fig. 7

figuring, laying out, etc., a long conical point should be used. A 2 H pencil will be found satisfactory. For the drawing itself, one 4 H pencil and one 6 H pencil, carefully sharpened, are needed. After removing the wood, Fig. 4, the lead is made slightly conical, Fig. 5, and then formed as in Fig. 6, using fine sandpaper or a file. Fig. 7 shows enlarged side and front views of the lead.

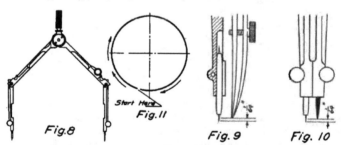

Fig. 8   Fig. 11   Fig. 9   Fig. 10

Use of the Compasses. — The compasses, Fig. 8, are used for drawing large circles. The needle point should be adjusted with the shoulder downward and so that the point extends about $1/_{64}$ inch beyond the pen point, Fig. 9. A 4 H or 6 H lead should then be sharpened as for the drawing pencil, and placed in the pencil leg. Remove the pen point from the compasses, insert the pencil leg, and fasten it. Then adjust the lead so that the end of it is about $1/_{64}$ inch above the needle point, Fig. 10. The joints in the legs are for the purpose of keeping the point and pencil perpendicular to the paper. The compasses should be operated with one hand (the right hand). The needle point should be placed in the center, and the marking point revolved clockwise. Once around is enough, starting at the point indicated in Fig. 11.

The bow instruments are used for small circles and divisions. The method of setting the points and using is the same as for the large compasses and dividers.

Use of the Dividers. — The dividers are used for transferring distances and for dividing lines. They should be handled with

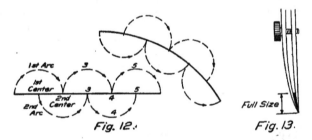

Fig. 12.   Fig. 13.

the right hand. When dividing a line, the points should be revolved in alternate directions, as indicated in Fig. 12. To divide a line into three parts, first set the dividers at a distance estimated to be about one third. Try it, and if too short, increase the distance between the divider points by one third of the remaining distance. If too long, decrease the distance between the divider points by one third of the distance which they extend beyond the end of the line. Repeat the operation if necessary.

The Use of the Ruling Pen. — The ruling pen is used for inking the straight lines, after the pencil drawing is finished. Ink is placed between the nibs of the pen by means of a quill which is attached to the ink bottle stopper. Care should be taken to prevent any ink from getting on the outside of the pen. The proper amount of ink is shown in Fig. 13. The pen should be held in a vertical position, and guided by the T square or triangle. It may be inclined slightly in the direction of the line which is being drawn, but the point must always be kept from the angle formed by the paper and the guide. Do not hold the pen too tightly, or press against the guide. Both nibs of the pen must touch the paper. Frequent cleaning of the pen is necessary to obtain good lines. The same methods apply to the compass and bow pens.

Character of Lines. — All pencil lines should be fine, clear, and sharp. For most purposes continuous pencil lines may be used. The character and weight of ink lines for use on drawings, may be found by reference to Fig. 14.

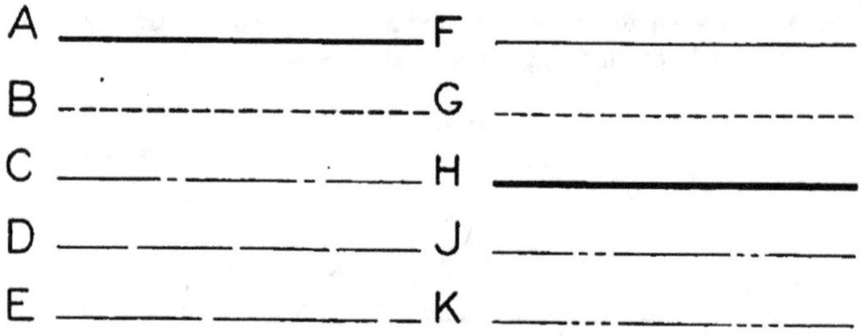

Fig. 14.

A. Full line for representing visible surfaces.

B. Dotted line used with A for representing invisible surfaces. Dots about $1/16$ inch long and very close together.

C. Center line — very fine dot and dash.

D. Witness line — short dashes.

E. Dimension line — long dashes, or fine full line. D and E are often made the same.

F. Fine line for shaded drawings.

G. Dotted line for shaded drawings.

H. Shade line for shaded drawings, about three times thickness of fine line.

J. and K. for special purposes, representing conditions not specified above.

When shade lines are not used, a fairly wide line should be adopted as wearing better, and giving better blueprints. The width of line will depend somewhat upon the drawing. Large simple drawings require a wide line, while small intricate drawings necessitate narrower lines. Drawings which are large and still have considerable detail in parts require more than one width of line. An experienced draftsman will use wide lines for the large and simple parts, reducing them for the complicated places in such manner that the different widths of lines are not noticeable. The student is cautioned to proceed slowly and strive for a uniform width of line until experience teaches discretion.

Center lines are drawn very fine, and are composed of dots and dashes. All symmetrical pieces should have a center line. All circles should have both horizontal and vertical center lines.

Much information concerning the many different kinds of drawing devices used by draftsmen for saving time and other purposes can be found in the catalogues of drawing material companies.

CHAPTER II

LETTERING

Lettering. — The subject of lettering in connection with working drawings is of great importance. Neat, legible letters, made free hand and with fair speed, are required. This chapter will deal with such letters. Those who wish to pursue the subject further should procure a good book on lettering, such as French & Meiklejohn's "Essentials of Lettering," published by McGraw-Hill Company, New York, or Daniels' "Freehand Lettering," published by D. C. Heath & Company, Boston, Mass. Either of these books may be obtained for $1.00.

Great care and continual practice are necessary to do good lettering, but the appearance of neatness, the greater ease of reading, and lessened liability of mistakes, make up for the extra time and work.

Commercial Gothic Letters. — Commercial gothic and lower case letters or small letters are the forms most used by engineers and draftsmen. These are shown in Fig. 15, with the proportions and directions for drawing the various lines. The vertical capitals and lower case letters are shown in Fig. 16. The same proportions and order of strokes apply to the vertical letters. The inclined letters should have a slope of about 3 to 8, as shown in Fig. 17. Some draftsmen use the 60° slope, but this does not give as pleasing a letter (Fig. 18).

In all cases very light pencil guide lines should be drawn to limit the tops and bottoms of the letters. The size of the letters is determined to some extent by the character of the work, but for most drawings the capitals should be $1/8$ inch high, and the small letters about two thirds as high (Fig. 17). For penciling use a 2 H pencil, with a well sharpened round point. For inking, a ball point pen may be used for fairly large letters, and Gillotts 404 or 303 for small letters. The pen may be dipped into the ink and the surplus shaken back into the bottle, or the quill may be used as with the ruling pen. For good work, the pen

7

Fig. 15

point must be kept clean, requiring frequent wiping. The pen point should be kept pointed toward the top of the paper.

Proportions and Forms. — The proportions and shapes of the various letters should be studied and drawn to a large scale. For purposes of study, the letters are divided into groups. The following points should be observed. Rounded letters, such as

I L T H F E N M

Z Y A K V W X

O Q C G U J D

B P R S I 2 3

4 5 6 7 8 9 &

a b c d e f g h i j

k l m n o p q r s

t u v w x y z

Fig. 16

C, J, O, Q, and S, may extend very slightly outside of the limiting lines. Pointed letters, like A, V, and W, may have the point extending very slightly above or below the guide lines. The horizontal bar in the letters B, E, F, H, and R is very slightly above the middle, and for the letter P it is very slightly below the middle of the vertical height. For the letter *A* the bar is placed about one third the height of the letter. The letter *W* is wider than it is high. The two outside strokes of the *M* are parallel.

Letter Spacing. — The spacing between the letters when combined to form words will vary with different arrangements. The only general rule which can be given is that the area between

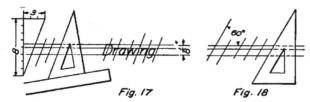

Fig. 17   Fig. 18

the letters should be about equal. A few illustrations will be given, showing the positions of some combinations of letters. When such letters as *A* and *T*, or *A* and *V*, are used, they should in general be placed close together, as in Fig. 19. A few words are shown in Fig. 20. In the lines marked WRONG the letters are equal distances apart. In the lines marked RIGHT the letters are spaced so that the areas between them are about equal. The

Fig. 19

combination of letters in each word, or the combination of words in a line, will determine the spacing of the letters.

Titles. — The matter of titles for drawings is subject to great variation. The titles for detail drawings may or may not contain the name and location of the concern. The name of the machine, its size and number, the names of the details, scale, date, and names or initials of the draftsman and engineer, should be given. An example is shown in Fig. 21. Assembly drawings generally have more elaborate titles. Good titles cannot be made by rule, though a few suggestions may be of assistance. It is often advisable to center the lines composing the title. This

Fig. 20

may be done by counting from each end of each line toward the center, and placing the middle letter or space on the center line. The line can then be completed by working in opposite directions from the center line. The important facts should be given due

---

500 GALLON
STEAM JACKETED KETTLE

SCALE 1½"=1 Foot	APPROVED BY	
DRAWN BY D.F.A.	DATE Sept. 12, 1917	
TRACED BY W. E.	ORDER·NO. B-462	
CHECKED BY C.S.	REVISED Jan. 4, 1919	

Draw. No. 4-C-145

Fig. 21

prominence. This may be done by using large letters, by using heavier or blacker letters, by wide spacing between letters, or by using extended letters. The element of time should be considered in the selection of letters. In general, the title should be placed in the lower right-hand corner of the drawing, and

OHIO TECHNICAL DRAWING SCHOOL COLUMBUS, OHIO.	CYLINDER FOR 4×5 VERT. ENGINE Full Size	Drawn by J. K. Traced by J. K. Approved by L. J. Date Sept. 12, 1917

Fig. 22

may or may not be "boxed in" (Fig. 21). Some concerns use a title extending across the end of the drawing, in which case it forms a "record strip" (Fig. 22).

Bill of Material. — A bill of material is often put upon each detail drawing in connection with the title. Sometimes a separate

PART No	NAME OF PART	No WANTED	MATERIAL	PATTERN No.
1	Spindle	1	Steel	—
2	Spur Center	1	Steel	—
3	Cap Screw	4	Steel	—
4	Cone Set Screw	1	Steel	—
5	Box Pins	2	Steel	—
6	Thrust Screw	1	Steel	—
7	Clamp Screw	1	Steel	—
8	Box	2	Brass	K-45
9	Washer	1	Brass	W-133
10	Thrust Check Nut	1	Steel	—

Fig. 23

sheet is made containing a list of drawings, material, number required, pattern number, location, etc., for the entire machine or construction. Both methods may be used together. The advantages of a separate list are apparent in certain classes of machines where some drawings are used for many different machines. Bolts, pins, keys, and similar small parts are often given a number, which is used to designate them. The application and uses of lists are so varied that they must be learned for the company where one is employed. A material list is shown in Fig. 23.

CHAPTER III

CONSTRUCTIONS

Essential Constructions. — Geometry forms the basis of the constructions used in the making of drawings. A knowledge of some of the principles of geometry is therefore essential. A point indicates position in space. When a point is moved it

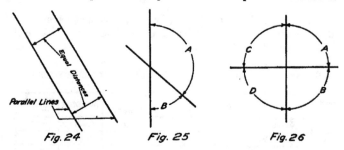

Fig. 24   Fig. 25   Fig. 26

generates a line, which may be either straight or curved. A surface may be formed by moving a line. A plane surface is one which will contain two intersecting straight lines. Two straight lines are said to be parallel when they are everywhere the same distance apart (Fig. 24).

Angles. — When two lines cross they form angles. The size of the angle is determined by the amount of opening between the lines. The angle A, in Fig. 25, is greater than the angle B. If the lines are revolved about their intersection, so that the angle A is made equal to the angle B, then both angles are called "right" angles (Fig. 26). The angles C and D are also right angles, so that all four angles are equal. As shown, each angle is one fourth the way around the point of intersection. When a right

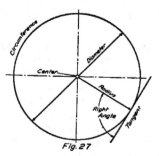

Fig. 27

13

angle is divided into 90 equal small angles, each of these small angles is called a "degree." Then it takes 4 times 90, or 360, degrees to go all the way around the point where the lines cross.

Circles. — A circle is a curved line formed by moving one point around another point and at a constant distance from it (Fig. 27). The curved line is called the circumference. The constant distance is called the radius, and the fixed point is called

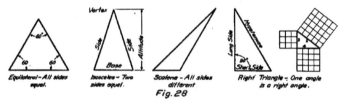

Equilateral - All sides equal. Isosceles - Two sides equal. Scalene - All sides different. Right Triangle - One angle is a right angle.

Fig. 28

the center. Lines drawn from the circumference to the center form angles, which are measured in degrees. Two lines crossing each other at the center of the circle, and making equal angles with each other, form four right angles, so that a circle is said to contain 360 degrees (written 360°). A piece of the circumference is called an arc. Other features of a circle are indicated in

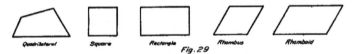

Quadrilateral Square Rectangle Rhombus Rhomboid

Fig. 29

Fig. 27. The length of the circumference is equal to the diameter times 3.1416. (3.1416 is called "pi," and is often written π.)

Plane Figures. — A plane figure made up of three lines is called a triangle. There are several kinds of triangles (Fig. 28). All three angles of any kind of a triangle, when added together, are always equal to 180 degrees. The sides of the right triangle have a very useful relation to each other, which is illustrated in Fig. 28. If the length of each of the two sides is squared and added together, the sum will be equal to the square of the length of the hypotenuse; thus, in the figure,

$$(3)^2 + (4)^2 = (5)^2$$

or

$$9 + 16 = 25$$

A plane figure made up of four lines is called a quadrilateral.
When the opposite sides are equal and parallel, the figure is
called a parallelogram.   There are several kinds (Fig. 29.)

Other regular plain figures are shown in Fig. 30.

Solids may have almost any form.   The names and appearances of a number of solids are shown in Fig. 31.

There are many geometrical constructions which are of use in

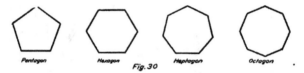

Pentagon   Hexagon   Heptagon   Octagon
Fig. 30

mechanical drawing.   Detailed instructions for the solution of
some of these problems follow.   These problems should be studied
carefully and be fully understood.   They should be worked out
with a very sharp pencil, fine lines, and extreme accuracy.

To bisect a Line. — (To divide a line into two equal parts):
Given the line AB (Fig. 32).   Using points A and B as centers,

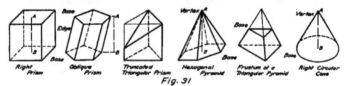

Right   Oblique   Truncated   Hexagonal   Frustum of a   Right Circular
Prism   Prism   Triangular Prism   Pyramid   Triangular Pyramid   Cone
Fig. 31

and a radius greater than one half the length of the line, draw
the arcs 1 and 2.   Through the points where these arcs cross
each other, draw the line CD, which will divide the line AB into
two equal parts.   The lines CD and AB form right angles, and
are said to be perpendicular to each other.   The steps used in
solving this problem are illustrated in Fig. 32.   Any given line
is shown at a.   At b is shown the given line and the arc of a circle
having a radius greater than one half the line, and its center at
the upper end of the line.   At c another arc has been drawn,
having the same radius as before, but with its center at the lower
end of the line.   At d a line has been drawn through the intersections of the two arcs, dividing the given line into two equal
parts.   It is not necessary to draw the whole of the arcs or the

intersecting lines. The usual appearance of the completed problem is shown at *e*.

To bisect an Angle (Fig. 33). — Given the angle *AOB*. With *O* as a center, and any radius, draw an arc intersecting the sides of the angle in points 1 and 2. With points 1 and 2 as centers, and a constant radius, draw arcs cutting each other at *C*. The line *OC* will bisect the angle.

To divide a Line into Any Number of Equal Parts (Fig. 34). —

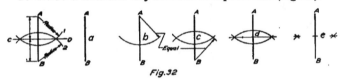

Fig. 32

Given the line 5′*B*. It is required to divide the line into five equal parts. From one end of the line draw another line, making an angle with it, such as *B*5. On *B*5, using any convenient setting of the dividers, step off five equal spaces. Join the end of the last space with the end of the given line. Through points 4, 3, 2, and 1 draw lines parallel to 5, 5′, intersecting the given

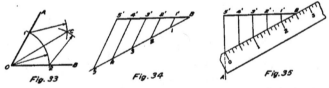

Fig. 33   Fig. 34   Fig. 35

line at points 4′, 3′, 2′, and 1′. The line 5′*B* will then be divided into five equal parts.

Another method of dividing a line is illustrated in Fig. 35. From one end of the given line draw a perpendicular such as 5′*A*, using the triangle and T square. Next place the scale in such a position that one end of any five equal divisions is at point *B*, and the other is on the line 5′*A*. Mark opposite each of the divisions, and through each mark draw a vertical line intersecting the given line, which will then be divided into five equal parts.

To copy an Angle (Fig. 36). — Given the angle *AOB*. To construct another angle equal to it. Draw a line *A′O′*. With *O* as a center and any radius, draw an arc cutting the sides of the angle at 1 and C. With O' as a center, and the same radius, draw the arc $1'C'$. With $1'$ as a center, and a radius equal to the chord $1C$, draw an arc cutting the arc $1'C'$ at C'. Draw $C'O'$. Angle $A'O'B'$ will then be equal to the angle AOB.

To construct a Triangle, having given the Three Sides (Fig. 37). — Given the three lines A, B, C. Draw line A', equal to line

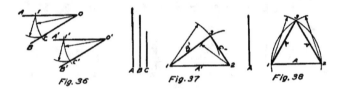

Fig. 36   Fig. 37   Fig. 38

A. With 1 as a center, and a radius equal to line B, draw an arc. With point 2 as a center, and a radius equal to line C, draw another arc, cutting the first arc at point 3. Join point 3 with points 1 and 2, completing the required triangle.

To construct an Equilateral Triangle (Fig. 38). — Given one side of the triangle, A. Draw line 1–2, equal in length to line A.

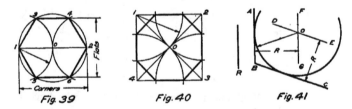

Fig. 39   Fig. 40   Fig. 41

With 1 and 2 as centers, and radius equal to line A, draw arcs intersecting at 3. Join point 3 with points 1 and 2, completing the required triangle.

To construct a Regular Hexagon (Fig. 39). — If the distance across corners is given, draw a circle having a radius equal to one half this distance. Draw the diameter 1O2. With points 1 and 2 as centers, and the same radius, draw arcs cutting the circle at points 3, 5, 4, and 6. Join these points to complete the required hexagon. It will be noted that the radius used as a chord divides the circumference into six equal parts. The 30 × 60 triangle may be used to construct a hexagon. Explain how.

To construct a Regular Octagon (Fig. 40). — Given the square 1–2–3–4. With the corners of the square as centers, and a radius equal to one half the diagonal, draw arcs cutting the sides of the square. Join the points thus found, completing the required octagon. An octagon may be constructed inside of a circle by using the 45-degree triangle. Explain how.

To draw an Arc of a Circle, having a Given Radius, and tangent to Two Given Lines (Fig 41). — Given the lines AB and BC and

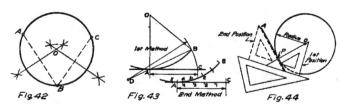

Fig. 42   Fig. 43   Fig. 44

the radius R. Draw DE parallel to BC, and at a distance equal to R from it. Draw FG parallel to AB, and at a distance equal to R from it. Where DE and FG cross, gives point O, the center of the required arc.

To draw a Circle, Passing through Any Three Points (not in the Same Straight Line) (Fig. 42). — Given points A, B, and C. Draw lines AB and BC. Bisect lines AB and BC, using the construction of Fig. 32. Where the bisecting lines cross at O is the center of the required circle. The radius is the distance from O to any of the three points.

To find the Length of an Arc of a Circle, and measure it on a Straight Line (Fig. 43). — *First method* (when angle AOB is less than 60 degrees): Given arc AB with center at O. From one end of the arc draw the tangent line AC. Draw line AB and extend it to D, making AD equal to one half of line AB. With D as a center, and radius DB, draw arc BC. Then AC will be a straight line equal in length to the arc AB. *Second method:* Draw tangent AC as before. Set the dividers at a small distance. Start at point B, and space off the points 1, 2, 3, etc., along the arc, until point 5 is reached. (Point 5 may come at any place near the point A.) Do not remove the dividers from the paper. Step back along the line the same number of spaces, as shown. The line AC will then be very close to the length of the arc AB. By taking small spaces, the chords may be assumed equal to the arcs.

To draw a Tangent to a Circle at a Given Point on the Circle (Fig. 44).—Given point P. Place one triangle with its hypotenuse passing through the given point, and the center of the circle as indicated in first position. Using the other triangle as a base, turn the first triangle over into the second position, and move it until its hypotenuse passes through point P, when the tangent

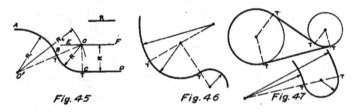

Fig. 45   Fig. 46   Fig. 47

AP may be drawn. The base triangle must be held firmly in place in the one position.

To draw the Arc of a Circle of Given Radius, tangent to an Arc and a Straight Line (Fig. 45).—Given arc AB, line CD and radius R. Draw line EF parallel to CD, and at a distance R from it. With radius $R_2 = R_1 + R$ and center O', draw an arc cutting line EF at O, the center of the required tangent arc. Note the points of tangency, which are marked T. The point of tangency of any two arcs is always on the line joining their centers. This is further illustrated in Figs. 46 and 47, where the points of tangency are marked T.

The Ellipse.—An ellipse (Fig. 48), is a curve formed by a point moving so that the sum of its distances from two fixed points is a constant. Each of the two points F_1 and F_2 is called a focus. The

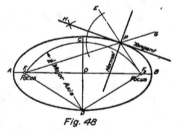

Fig. 48

longest line, AB, drawn through the center is called the major axis. The shortest line, CD, is called the minor axis. The constant distance is equal to the major axis. A tangent to an ellipse at any point may be constructed by drawing lines from the point to the foci. Extend the lines and bisect the angle F_1PE, or the angle F_2PG. This bisecting line PH is the required tangent. A line through the point P and perpendicular to the tangent is called a normal. The major and minor axes

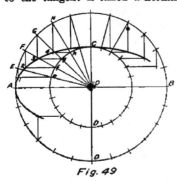

Fig. 49

of an ellipse being given, the foci may be located by drawing an arc with C or D as a center, and a radius equal to one half of the major axis. This arc will cut the major axis at the foci.

To draw an Ellipse by the Concentric Circle Method (Fig. 49.) — Given the major and minor axes AB and CD. With O as a center, draw circles having the major and minor axes as diameters. Draw radial lines OeE, OfF, etc., dividing the circles into a number of parts. Where the radial lines cut the large circle, draw perpendicular lines. Where the radial lines cut the small circle, draw horizontal lines. The intersection of a vertical and horizontal line from the same radial line will determine a point on the ellipse, as indicated at 1, 2, 3, and 4. Determine as many points as necessary, and draw the curve through them very lightly free hand. It may then be strengthened, using an irregular curve.

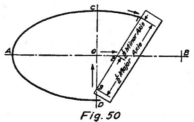

Fig. 50

To draw an Ellipse by the Trammel Method (Fig. 50). — Given the major and minor axes AB and CD. On a small strip of paper mark off one half the minor axis and one half the major axis, as shown in the figure. Place the point 3 on the minor axis and the point 2 on the major axis. Make a mark on the paper opposite the point 1. Move the point 3 along the minor axis, keeping the point 2 on the major axis and moving it as indicated by the arrows. The point 1 will then trace out the required ellipse. The usual method is to place the trammel in a number of positions, and make marks on the paper opposite the successive positions of point 1.

To draw a Curve having the Appearance of an Ellipse, by means of Circular Arcs (Fig. 51). — Given the major and minor axes AB and CD. On the minor axis lay off $O3$ and $O1$, each equal to the difference between the major and the minor axis. On the major axis lay off $O2$ and $O4$, each equal to three fourths of $O3$. With point 1 as a center, and a radius equal to $1C$, draw the arc ECG.

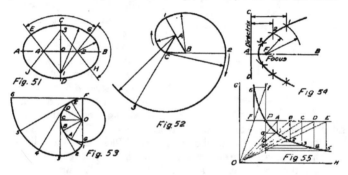

Fig. 51   Fig. 52   Fig. 53   Fig 54   Fig 55

With 3 as a center, and the same radius, draw the arc JDH. With 2 and 4 as centers, and a radius equal to $2B$, draw the arcs GBH and EAJ.

The Involute. — Tie a piece of string about a lead pencil point. Place the triangular scale with its end resting upon a piece of paper. Wind the string about the scale, keeping the pencil point toward the paper. Hold the scale firmly with one hand. Keeping the string tight, and the pencil point on the paper, unwind from the scale. The curve thus formed is the involute of a triangle. The involute of any other figure may be obtained by unwinding a string from the desired form.

To draw the Involute of a Triangle (Fig. 52). — With A as a center, and AC as a radius, draw an arc until it reaches the extension of side AB at point 1. With point B as a center, and $1B$ as a radius, draw an arc from 1 until it reaches the extension of side CB at point 2. The curve may be continued by increasing the radius each time that it passes the extension of one of the sides. Compare this curve with the one drawn by means of the triangular scale and string.

To draw the Involute of a Circle (Fig. 53). — Divide the arc of a circle into a number of equal parts. Draw the radial lines OA, OB, etc. At the end of each radial line draw a tangent. Starting at point A, lay off the distance $A1$ on the tangent equal to the arc AG, using the second method of Fig. 43. Starting at B, lay off the distance $B2$ on the tangent, equal to the arc BAG. Continuing, lay off on each tangent a distance from the point of tangency equal in length to the arc of the circle, measured from the point of tangency to the point G.

The Parabola. — A parabola is a curve formed by a point moving so that its distance from a line called the directrix is always equal to its distance from a point called the focus (Fig. 54). To draw a parabola, having given the directrix CAD, and the focus F. Draw a line parallel to the directrix, and at any distance from it. Using this distance as a radius, and F as a center, draw an arc, cutting the parallel line at point 1. Draw as many lines as may be necessary, parallel to the directrix, and using their distances from the directrix as radii, with F as a center, draw arcs cutting them, thus locating points on the required parabola.

To draw an Equilateral Hyperbola (Fig. 55). — Given the point P and the axes GO and OH. Draw horizontal and vertical lines through point P. On each side of point P step off equal distances PF, PA, AB, etc. Draw lines from O to each of the points thus determined. Where line OA crosses the vertical line at point a, draw a horizontal line. Through point A draw a vertical line intersecting the horizontal line just drawn at point 1, a point on the required curve. Horizontal and vertical lines drawn from the diagonals will locate other points on the curve, as shown at 2, 3, 4, 5, and 6.

CHAPTER IV

PROJECTIONS

Purpose of Drawings. — The representation of objects and constructions having three dimensions upon a surface having two dimensions has been accomplished in many ways, some of which are illustrated in Fig. 56.

Drawings have two principal uses which are:

    I. To tell the shape,
    II. To tell the size.

A drawing tells the shape by the position of the various lines,

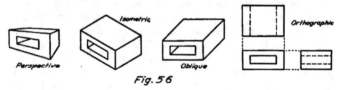

Fig. 56

while numbers are used to tell the size. These numbers are called dimensions.

Orthographic Projection. — Most engineering drawings are made in "orthographic projection." By this means the shape and proportions of a construction may be accurately defined. The number of views depends upon the object or construction to be described. This can be understood by reference to Fig. 57, which shows two views of a cylinder. The upper view shows the circular form and the lower view shows the height of the cylinder. Notice that the diameter of the upper view is the width of the lower view, and that the two views are included between parallel vertical lines. An object requiring three views is shown on Fig. 58. Note the arrangement of the

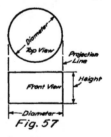

Fig. 57

23

views. The top and front views are included between parallel vertical lines, and the front and side views are included between parallel horizontal lines.

The Planes of Projection. — The method of obtaining the views and getting them in the correct relative positions will be

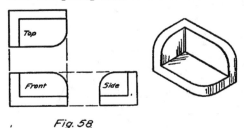

explained in connection with Figs. 59 and 60. Consider two glass planes, one horizontal and one vertical (Fig. 59), with an object placed in the angle thus formed. By

Fig. 58.

looking through the vertical plane the front of the object may be seen, and if this view is marked out on the glass, it is called the front view, elevation, or vertical projection. If, instead of look-

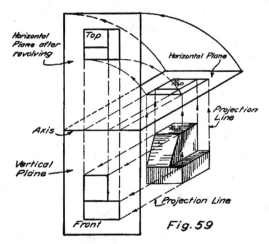

Fig. 59

ing through the glass, we consider that lines have been drawn from every point in the object perpendicular to the vertical plane, the object is said to be projected out to the vertical plane. The lines are called projection lines. By joining the points in which the projection lines touch the vertical plane the front view will be obtained. In the same manner the top view, plan, or horizontal projection may be found by projecting up

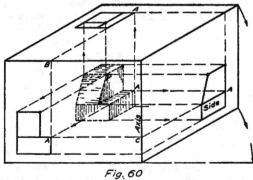

Fig. 60

to the horizontal plane. If the joint between the two planes is now taken as an axis, the horizontal plane may be revolved up about the axis until it is in the same plane with the vertical plane. This brings the top view directly above the front view.

By placing a third glass plane at one side of the object, and perpendicular to the other two planes, as shown in Fig. 60, a side view may be obtained. The plane containing this side view can be revolved about the axis shown until

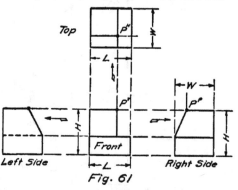

Fig. 61

it is in the same plane with the vertical plane. This brings the side view on the same line with the front view, as shown in Fig. 61, where the three views are in their correct positions.

Some Rules. — The following points should be thoroughly understood, as projection is the basis for all shop drawings.

Note the three views of the point P in Figs. 59, 60, and 61. Locate other points in the same manner until all points on the object are accounted for in each of the three views.

Horizontal distances (as L) show the same in the top and front views.

Vertical distances (as H) show the same in the front and side views.

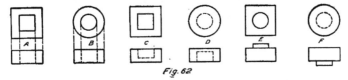

Fig. 62

Vertical distances (as W) in top view are horizontal distances (W) in the side view.

The top view is the same length as the front view.

The top view is the same width as the side view.

The front view is the same height as the side view.

The front of the side view is toward the front view.

The front of the top view is toward the front view.

The arrows (Fig. 61) indicate the relation of the front view to the other views.

Note the difference between the left side view and the right side view.

Lines which represent visible surfaces are full lines.

Lines which represent invisible surfaces are dotted lines. (See left side view, where it is necessary to look through the object in order to locate the horizontal dotted line.)

The top and front views of any point are always in the same perpendicular line.

The front and side views of any point are always in the same horizontal line.

Dotted Lines. — The question of dotted lines is illustrated in Fig. 62, where two views of several objects are shown. A is a square prism with a square hole all the way through; B is a cylinder with a circular hole all the way through; C is a square prism with a square hole extending from the top down to the depth shown in the front view (note that the top views of A and C are the same, and that the front views show the extent of the holes); *D* is a cylinder with a hole extending up part way from the bottom, as shown in the front view, therefore the hole shows dotted in the top view; *E* is a square prism with a cylindrical boss on top; *F* is a cylinder with a smaller cylinder extending downward from the under side, thus the small cylinder is dotted in the top view; compare *F* and *D*, which show that it is necessary to read both views to determine the object. A large number of all sorts of projection problems should be solved to obtain a thorough understanding of orthographic projection.

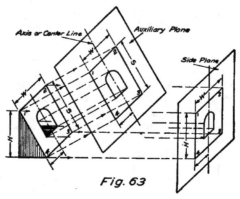

Fig. 63

Auxiliary Views. — The three planes just described are perpendicular to each other, like the boards coming together at the corner of a box. The faces of an object which are parallel to the three planes are projected to these planes in their true size and shape. It is often desirable to show the true shape of a face which is not parallel to any of the three regular planes. In such cases, Fig. 63, an extra plane called an auxiliary plane may be used. This extra plane is placed so as to be parallel to the inclined face. The inclined face is then projected to the auxiliary plane by perpendicular projecting lines, as shown in Fig. 63.

The distances *W* and *S* then show in their true length and the hole shows the true shape in which it cuts the inclined face. Compare the auxiliary plane with the side plane. Notice that the distance *W* shows in its true length in the side plane, but that the vertical dimension is *H*, which is shorter than *S*. The auxiliary plane and the side plane may be revolved about the center line or axis until they are parallel to the plane of the paper. This has been done in Fig. 64, where the object is shown by its projections. Note the location of the points 1, 2, 3, and 4. The center

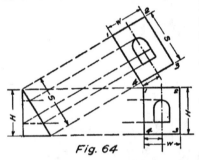

line of the auxiliary view is parallel to the inclined face. The width W is the same in the auxiliary view and in the side view. The points 1, 2, 3, and 4 are located in the auxiliary view by projecting lines perpendicular to the inclined face which cross the center line at right angles. The distances on either

Fig. 64

side of the center line are then obtained from the side view and measured on the corresponding projection lines of the auxiliary view, as illustrated for point 4.

Compare Figs. 63 and 64 carefully.

Required Views. — A bracket is shown in pictorial form in Fig. 65, together with its three views in orthographic projection. Note the relation of the views. A picture of an object is shown

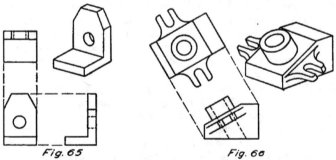

Fig. 65   *Fig. 66*

in Fig. 66. Since most of its detail is inclined, a side view and auxiliary view are used. In this way true shapes are shown. Other views are not needed. They would be somewhat difficult to draw and would not add anything to what is already shown. Very good practice is to be had by deciding the number of views and proper treatment for such machine parts and constructions as one encounters.

The Imaginary Cutting Plane. — It is not always possible to indicate easily and clearly the interior construction of a machine

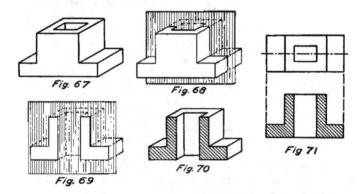

Fig. 67   Fig. 68   Fig. 69   Fig. 70   Fig 71

or part by means of dotted lines. In such cases resort is had to imaginary cutting planes which reveal the hidden parts.

Such an imaginary cutting plane passing through the object of Fig. 67 is shown in Fig. 68. The part in front of the cutting plane is removed in Fig. 69, leaving the object as shown in Fig. 70, where the surfaces cut by the plane are indicated by parallel inclined lines. Such a surface is said to be cross-hatched or section-lined. The view is called a section, or sectional elevation. In orthographic projection the two views are drawn as in Fig. 71, where the section occupies the same position relative to the top view as the front view which it replaces. Note that the top view is shown complete. The top edge of the cutting plane is shown as a center line in the top view. The rules of projection apply to sectional views. The object is imagined to be cut by a plane and the part in front of the plane removed in order to show the cut surfaces and the details beyond the cutting plane.

Representation of Cut Surfaces. — The surfaces which lie in the imaginary plane are indicated by a series of parallel lines. Different pieces are shown by changing the direction of the lines. The width of spacing for section lines is determined by the area to be sectioned. Different materials are sometimes indicated by different forms of section lining. Fig. 72 gives the forms suggested by a committee of the American Society of Mechanical

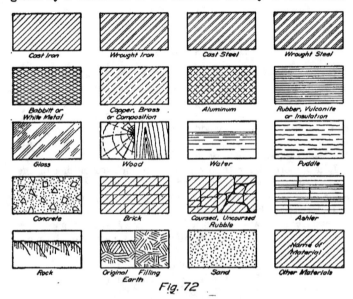

Fig. 72

Engineers. The character of sectioning should not be depended upon to tell the material. It should always be given in a note if it is not perfectly evident.

CHAPTER V

MATERIALS AND STRESSES

Materials. — Engineering constructions must carry loads and transmit motions. For such purposes various materials are made use of according to their adaptability. The most used material is iron in its different forms — cast iron; wrought iron; steel; and the steel alloys. In addition to iron there are the yellow metals, or brass and bronze compositions, white metals or babbitt, tin, lead, etc., and the various timbers.

It is important for the draftsman to know something of the properties of these materials, the methods of forming into machine parts, and the relative expense, so that a proper selection of material may be made for the particular case in hand.

Cast Iron. — Cast iron is a hard, brittle, granular substance obtained by burning the impurities from various ores. the most common being

> Magnetic Oxide, or Magnetite
> Ferric Oxide, or Red Hematite
> Brown Hematite
> Spathic Ore

Cast iron contains carbon and various impurities, such as silicon, manganese, phosphorus, sulphur, etc.

White Iron. — There are two principal kinds of cast iron: white cast iron, in which the carbon is chemically combined, and gray cast iron, in which the carbon is free or mixed in the form of graphite. White cast iron contains a small amount of carbon, and is very hard and brittle. It is used in the manufacture of wrought iron and steel. White cast iron is very difficult to machine.

Gray Iron. — Gray cast iron contains some carbon in chemical combination and a considerable amount in the form of graphite, which is mixed with the iron. Gray iron is softer than white iron and is easily machined. It contains from 0.5 per cent to 1 per cent of combined carbon up to 2 per cent.

31

Properties of Cast Iron. — Cast iron is the most useful of metals, as it can be readily melted and cast into any desired form by first making a mold. For this reason it is adapted for making complicated shapes. Its cheapness renders it available where

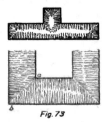

Fig. 73   Fig. 74

rigidity and weight are required. Cast iron cannot be welded and has very little elasticity, so that it is not adapted for use where shocks and sudden loads are to be cared for.

Cast iron has a crystalline structure, and when cooling the crystals form at right angles to the surface. Where square corners are encountered the arrangement is as indicated in Fig.

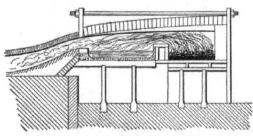

Fig. 75

73, in which fracture is likely to occur along *ab*, called the plane of fracture. This may be prevented by rounding, as in Fig. 74. Cast iron expands at the moment of solidifying, but shrinks upon cooling. This action sets up cooling strains in the casting, especially if there exists a considerable variation in the thickness of the section in the different parts of the piece. For this reason a uniform cooling arrangement is always desirable, and sudden changes in section should be avoided.

Cast iron is about four times as strong in compression as it is in tension.

Wrought Iron. — Wrought iron is almost pure iron, obtained by melting pig iron and squeezing out the impurities while it is in a plastic state. For such purposes a puddling furnace (Fig. 75) is used. Iron is put into the furnace and melted. When in a plastic state it is taken in the form of a ball on the end of a puddle bar and squeezed or pounded and heated again. This process is

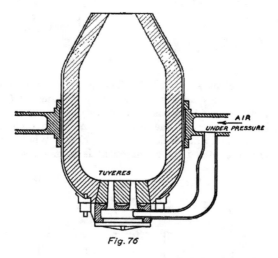

Fig. 76

continued until most of the impurities are burned or squeezed out. It is then rolled into bars or billets. These billets are further rolled into rods of various shapes and sizes called merchant bars. This rolling process gives the iron a fibrous structure due to a certain amount of impurities which remain after the puddling. Wrought iron contains a very small amount of carbon.

Properties of Wrought Iron. — Wrought iron is malleable and is the best material to withstand shocks. It stretches and so gives warning before breaking. It cannot be cast, but must be rolled or forged into the forms required. For this reason it is not adapted for complicated shapes. It can be welded, punched, bent, etc. Owing to its method of manufacture, it is expensive and is supplanted to a considerable extent by mild steel, which has a similar composition. Wrought iron is almost equally strong in tension and compression. It is stronger in the direction of the fibers than across them.

Steel. — Steel is made by burning carbon and impurities out of pig iron and then adding the desired amount of carbon. Another method is to add carbon to wrought iron. There are two processes of making steel from pig iron: the Bessemer process and the open-hearth process.

Bessemer Process. — In the Bessemer process from five to twenty tons of melted pig iron is put into a converter (Fig. 76). Air under a pressure of about twenty pounds per square inch is caused to pass in streams up through the metal, and the carbon and impurities are burned out. This requires about ten minutes, and leaves practically pure iron, to which the proper carbon content is added by putting in liquid spiegeleisen (white iron) or ferromanganese. This makes it into steel, which is poured into ingots. These ingots are rolled into blooms and other desired shapes.

Open-hearth Process. — By this process large amounts of steel are made at one time, generally about fifty tons. Steel, scrap, and pig iron are melted on the hearth of a Siemens regenerative furnace. The metal is kept in agitation by chemical reactions, caused by adding iron scale or scrap iron which furnish the necessary carbon.

Properties of Steel. — Steel is composed of iron and carbon in chemical combination. It has a uniform granular structure and may be formed to shape by forging, rolling, or casting. Steel varies greatly in its qualities, depending upon the carbon content. It is sometimes designated as

Soft Steel	about 0.19 % carbon
Medium Steel	” 0.30 % ”
Hard Steel	” 0.75 % up to 1.8 % carbon

Steel having less than 0.25 % is frequently called mild steel.

Malleable Iron. — Small parts of cast iron can be made less brittle by being surrounded by iron scale or some form of an oxide of iron and kept at a bright red heat for over sixty hours. In this way some of the carbon is removed and the material is made to resemble wrought iron. It is used for small pieces which cannot be easily forged. Hardware castings, pipe fittings, etc., are often made of malleable iron.

Suggestions for Selection of Material. — The best method of learning the proper materials to be used is by observation. The material best adapted cannot always be used, because of cost, method of shaping, etc. Ask *why*, when a special material is used. The "factor of cost" is always present — the "factor of safety" should always be considered first. Observe broken parts of machines as a valuable means of obtaining sound information. The use of special metals is often one of trial and observation. Some things which influence the selection of material are given below:

Method of Shaping

Casting	Cost of Pattern
Forging—Drop Forging	” ” Die
Pressing—Stamping	” ” ”
Extrusion—Drawing	” ” ”
Rolling	” ” ”

Number Required
Method of Finishing
Strength Required
Kind of Loads
Moving or Stationary Parts
Lightness or Weight
Wear
Where liquids or gases are used the chemical action must be considered.

Loads and Stresses. — The materials used in machines are subject to various loadings which must be resisted by these materials. The internal resistance must be equal to the external or applied load, or the part will fail. There are many ways of applying the load, each bringing into play a different form of resistance by the material. This resistance is called stress. Stress is a measure of the strength of the material to resist an external load. There are three kinds of simple stresses: tension, compression, and shear.

Axial Stresses. — A bar is a portion of material having a uniform section, such as a cylinder or prism. When a load is applied to a bar so as to be uniformly distributed it is called an axial load. Such a load produces a direct stress in the bar. The section made by passing a plane at right angles to the axis of the bar is called a cross section. The area of this section is the cross-sectional area and is usually spoken of as the area. It is generally measured in square inches.

When a load is applied to a bar so that it tends to lengthen the bar it produces a tensile stress (Fig. 77). When the applied load tends to shorten or compress the bar it produces a com-

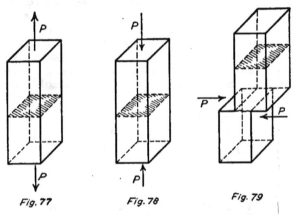

Fig. 77   Fig. 78   Fig. 79

pressive stress (Fig. 78). When the applied load acts at right angles to the bar and tends to push one cross-sectional plane by another it produces a shearing stress (Fig. 79).

Unit Stresses. — In order that the strength of various materials may be compared, the strength of a bar one inch square is used as a unit. The strength of such a bar is called the unit stress, or stress per square inch of cross-sectional area. The stress is usually given in pounds per square inch. To find the unit stress, divide the applied load by the cross sectional area, or:

Let A = area of cross section in square inches.

P = total load in pounds.

f = stress in pounds per square inch.

Then the unit stress is

$$f = \frac{P \text{ (load)}}{A \text{ (area)}}$$

Thus, if a rod has an area of $3^1/_2$ square inches and is subject to a load of 35,000 pounds, it has a unit stress of

$$f = \frac{P}{A} = \frac{35,000}{3.5} = 10,000 \text{ lb. per square inch.}$$

Elastic Limit. — From the formula given above it follows that if the load is doubled, the unit stress will also be doubled. This means that the unit stress is proportional to the load. By experiment it has been found that this law does not hold for all loads, but only up to a certain load (depending upon the material). This load or limit is called the elastic limit and is expressed in pounds per square inch. For stresses less than the elastic limit the increase or decrease in length of the bar is directly proportional to the stress. The increase or decrease in length is called the strain, and the total strain divided by the length is called the unit strain.

Let l = length in inches

e = change in length in inches

s = unit strain

Then

$$s = \frac{e}{l}$$

Modulus of Elasticity. — Below the elastic limit both the unit stress and the unit strain are proportional to the load, so that they bear a constant relation to each other. This relation is expressed as the quotient obtained by dividing the unit stress by the unit strain, which will give a constant called the modulus of elasticity and represented by E.

Then

$$E = \frac{\text{Stress}}{\text{Strain}} = \frac{f}{s}$$

Ultimate Strength. — The formula $f = \dfrac{P}{A}$ gives the unit stress of a material for a given load. If the load is sufficiently large the piece will break or rupture. The value of f when rupture takes place is called the ultimate strength of the material.

Factor of Safety. — The ultimate strengths of materials as well as the elastic limits are not constants, although most of them are pretty well known from large numbers of tests. However, it is not desirable to stress a material too near its elastic limit, as there may be imperfections or lack of uniformity. The manner in which the load is applied also affects the stress which it is safe to impose upon a given material. For this reason various "factors of safety" are used. A factor of safety is a number obtained by dividing the ultimate strength of a material by the unit stress actually imposed upon it. The actual stress is referred to as the safe working stress. Often the safe working stress is obtained by dividing the ultimate strength by a suitable factor of safety, depending upon the nature of the loading.

Average Values. — The values given in the following tables are averages and will serve for purposes of computation in the absence of more definite figures.

ELASTIC LIMITS

Material	Pounds per Square Inch	
	Tension	Compression
Cast Iron.............	6000	20,000
Wrought Iron.........	25,000	25,000
Steel.................	35,000	35,000

MODULI OF ELASTICITY

Material	Pounds per Square Inch
Cast Iron.............	15,000,000
Wrought Iron.........	27,000,000
Steel.................	30,000,000

ULTIMATE STRENGTHS

Material	Pounds per Square Inch		
	Tension	Compression	Shear
Cast Iron........	20,000	90,000	18,000
Wrought Iron.....	50,000	50,000	40,000
Steel.............	60,000 to 100,000	60,000 to 150,000	50,000 to 80,000

FACTORS OF SAFETY

Material	Dead Load	Live Load		
		One Kind of Stress	Alternate Tension and Compression	Varying Loads. Shocks
Cast Iron..........	4	6	10	15
Wrought Iron and Steel...	3	5	8	12

CHAPTER VI

SCREW THREADS

Uses of Screw Threads. — A screw is a cylindrical bar having a helical projection. The form of this helical projection varies, according to the uses to which the screw is put. Screws are used for the following purposes: To fasten parts of machines together; to transmit motion; to convert rotation into translation, or vice versa; for the adjustment of parts in their relation to one another.

The Helix. — A helix is a curve generated by a point moving equal distances lengthwise of a cylinder while it is moving equal

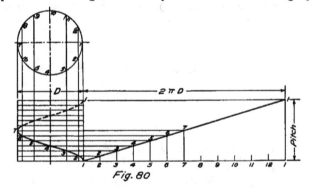

Fig. 80

distances around the cylinder. If a right triangle is wound around a cylinder the hypotenuse will form a helix. The points 1, 2, 3, 4, etc., of Fig. 80 will come at the same numbers on the curve when the triangle is wound around the cylinder. The pitch of a helix is the distance which the point moves parallel to the axis while it goes once around the cylinder.

To draw the Projections of a Helix. — In Fig. 80 let D be the diameter and let the pitch be the distance indicated. Divide the circle shown in the top view into any convenient number of equal parts, and draw vertical lines through each point. Divide the pitch into the same number of equal parts and draw horizontal

40

lines. For each space around the cylinder the point will move one of the spaces along the pitch, thus locating the curve as shown.

Parts of a Screw. — A screw is known by its outside diameter,

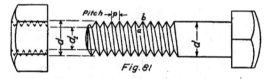

Fig. 81

indicated in Fig. 81 by d. The diameter d_1 is called the root diameter. Point b is the top of the thread and point a the bottom, or root. The area corresponding to d_1 is called the root area. One half the difference between the outside diameter and the root diameter is called the depth of the thread.

Right- and Left-hand Screws. — Screws may be either right- or left-hand. A right-hand screw thread (Fig. 93) requires the

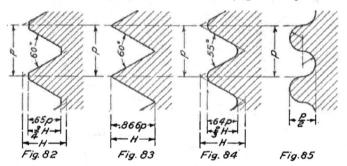

Fig. 82   Fig. 83   Fig. 84   Fig.85

screw to be turned in a clockwise direction to enter the nut. A left-hand screw thread (Fig. 99) must be turned counter-clockwise when entering. The pitch of a screw thread is the distance which the screw will advance for one complete turn for a single threaded screw.

Forms of Screw Threads. — The forms of screw threads are shown in the accompanying figures. Fig. 82 shows the Sellers, Franklin Institute, or U. S. Standard thread, as used quite generally in the United States. The proportions are indicated on the figures. The tops and bottoms of the V's are flattened so that the depth of the thread is decreased 0.25 the depth of the

V. The flats make the thread less liable to injury on the sharp
V's and less liable to weakening at the bottom of the grooves
than the sharper V thread shown in Fig. 83. This form of thread

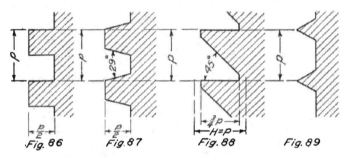

Fig. 86   Fig. 87   Fig. 88   Fig. 89

is also in quite general use. It is conveniently formed on a lathe,
and does not require a special tool, or regrinding of the tool, as
is the case for the U. S. Standard. The angle for both the above
forms is 60 degrees.

The Whitworth thread, or standard of Great Britain, is illustrated in Fig. 84. In this form the angle is 55 degrees. The threads
are rounded off at the top and bottom, making a strong shape.

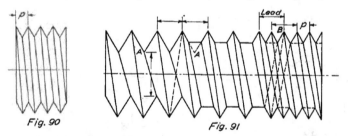

Fig. 90   Fig. 91

The forms described above are the ones most commonly used
for fastening parts together.

Fig. 86 shows the square thread, a form well adapted for use
in transmitting motion.

The Acme thread, a modification of the square thread, is shown
in Fig. 87. The angle may be either 29 or 30 degrees. This
form is used for transmitting motion. The relieving of the
thread allows the use of a split nut. A common example is the

lead screw of a lathe. Fig. 88 shows the buttress or breechlock thread so called from its use in guns to take the recoil. It is designed to take pressure in one direction only. This form has the strength in shear of the V form, but avoids the tendency to

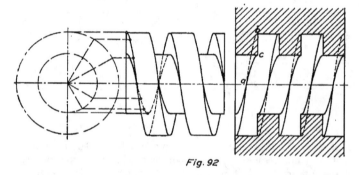

Fig. 92

split the nut. Fig. 85 shows the knuckle or rounded screw thread. This form can be cast in a mold. It is used only for rough work. Fig. 89 shows the common wood screw. An attempt is made to consider the differences in strength of the

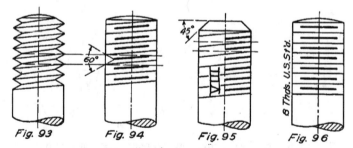

Fig. 93   Fig. 94   Fig. 95   Fig. 96

wood and steel. For adjustment, Figs. 82, 83, 84, 86, and 87 are used. Thrust screws for pillow blocks, crossheads, etc., are familiar examples of adjusting screws.

Multiple Threads. — Screws may have either single, double, or other multiple threads. A single-threaded screw consists of a single helical projection (Fig. 90). The pitch is the distance from one thread point to the next thread point. The lead is the distance which the screw will advance for one turn. When a large pitch is required on a small diameter, the arrangement of Fig. 91 would weaken the screw by reducing the root diameter (Fig. 91) at point *A*. To avoid this, two parallel helical projections may be used, as shown at *B* (Fig. 91). This is called a double

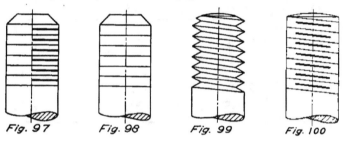

Fig. 97   Fig. 98   Fig. 99   Fig. 100

thread. Similarly, a triple or quadruple thread may be formed. In this manner a large lead may be obtained without lessening the strength of the screw.

Split Nut and Square Thread. — A portion of a square threaded screw, and a section of a nut for use with it are shown in Fig. 92.

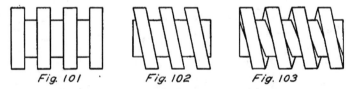

Fig. 101   Fig. 102   Fig. 103

The method of drawing the helix has already been explained. Note the dotted line *a-b*, which indicates the undercutting of surface *abc*, and shows why a split nut cannot be removed from a square threaded screw. The sloping side of the Acme thread does away with this undercutting and allows the removal of a split nut.

Conventional Representation of Screw Threads. — It is not often necessary to draw the helix in representing threads, as there are a number of conventional representations in use. Figs. 93 to 100 are common methods. Figs. 93 to 98 are for right-hand threads, Figs. 99 and 100 are for left-hand threads, and Figs. 96 to 98 are for either right- or left-hand threads. It is not

generally necessary to draw the pitch to scale. The distance
between lines may be estimated by eye and arranged to avoid
crowding of the lines. The number of threads per inch of other
than U. S. Standard should be given by note, as "12 *threads per*

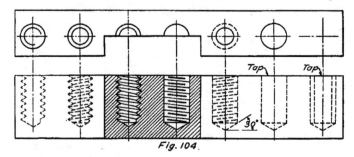

Fig. 104.

inch, right hand." This may be abbreviated to "12 *Thds. R. H."*
or "12 *Thds. L. H."* Sometimes the number is given for U. S.
Standard as indicated in Fig. 96, or the Roman numeral may be
used, as in Fig. 95.

Three representations for square threads are shown in Figs. 101,

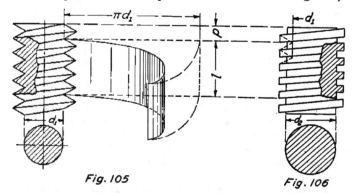

Fig. 105   Fig. 106

102, and 103. The square threads are generally drawn to scale,
and if of large diameter the helix may be drawn in, as in Fig. 92.

Threaded Holes. — Representations for threaded holes are
shown in plan, elevation, and section, in Fig. 104. It will be
observed that the lines representing the threads slope in the

opposite direction when the hole is shown in section. The reason for this is that the far side of the thread is seen. As shown, either single or double circles may be used in the plan view.

When the last two forms are used they should always be marked

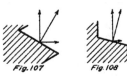

Fig. 107   Fig. 108   Fig. 109

"Tap" as indicated. For small diameters, the V's may be put in free hand. The lines representing the roots of the threads when visible are sometimes made heavier, but when dotted all lines should be of uniform thickness.

Strength of Screw Threads. — There are three methods of failure, shearing of threads, tension at the root of threads, and bursting of the nut.

Let f_s = unit shearing stress in pounds per square inch.

f_t = unit tensile stress in pounds per square inch.

p = pitch in inches

l = length in inches

The shearing strength of the V thread (Fig. 105) will be

$$P_s = \pi \, d_1 \, lf_s$$

and for square threads (Fig. 106) having the same outside diameter and pitch

$$P_s = \pi \, d_2 \, {}^1/_2 f_s$$

which shows that the square thread is much weaker in shear than the V thread.

The tensile strength of the V thread (Fig. 105) will be

$$P_t = {}^1/_4 \, \pi \, d_1^2 \, f_t$$

and for the square thread, Fig. 106, having the same outside diameter and pitch

$$P_t = {}^1/_4 \, \pi \, d_2^2 \, f_t$$

The V thread will have a considerable tendency to burst the nut, as shown in Fig. 107. As the angle between the threads decreases, this bursting tendency decreases until the square form is reached, when it becomes zero (Fig. 109).

The following tables give some desirable data concerning screw threads. Further information may be found in the handbooks published by Machinery and American Machinist.

DIMENSIONS OF U. S. STANDARD THREADS

Diameter	Threads per Inch	Diameter of Tap Drill	Root Diameter	Root Area
$1/4$	20	$3/16$	.185	.026
$5/16$	18	$1/4$	.241	.045
$3/8$	16	$5/16$	.294	.068
$7/16$	14	$23/64$	.345	.093
$1/2$	13	$13/32$	.400	.126
$9/16$	12	$15/32$	.454	.162
$5/8$	11	$17/32$	.507	.202
$3/4$	10	$5/8$	.620	.302
$7/8$	9	$3/4$	.731	.420
1	8	$27/32$	.838	.551
$1 1/8$	7	$31/32$	.940	.693
$1 1/4$	7	$1 3/32$	1.065	.889
$1 3/8$	6	$1 3/16$	1.159	1.054
$1 1/2$	6	$1 5/16$	1.284	1.293
$1 5/8$	$5 1/2$	$1 13/32$	1.389	1.515
$1 3/4$	5	$1 1/2$	1.491	1.744
$1 7/8$	5	$1 5/8$	1.616	2.049
2	$4 1/2$	$1 3/4$	1.711	2.300

TENSILE STRENGTH OF U. S. STANDARD SCREW THREADS

Diameter	Threads per Inch	Total Strength of One Bolt for Unit Stresses of		
		4000	5000	6000
$1/4$	20	105	135	160
$3/8$	16	270	340	405
$1/2$	13	500	625	750
$5/8$	11	805	1010	1210
$3/4$	10	1200	1500	1800
$7/8$	9	1680	2100	2520
$1 7/8$	8	2200	2750	3300
$1 1/8$	7	2770	3460	4160
$1 1/4$	7	3120	3900	4680
$1 3/8$	6	4240	5300	6360
$1 1/2$	6	5120	6400	7680
$1 5/8$	$5 1/2$	6120	7650	9180
$1 3/4$	5	7040	8800	10560
$1 7/8$	5	8120	10150	12180
2	$4 1/2$	9200	11500	13800

CHAPTER VII

BOLTS AND SCREWS

THE most common fastening for holding parts of machines together is some form of bolt or screw. There is a great variety of forms, many of which are shown in this chapter.

U. S. Standard Bolts. — Figs. 110 and 111 show the proportions of the U. S. Standard hexagonal bolt head and nut. As indicated, there are two general forms, chamferred (Fig. 110) and rounded (Fig. 111). The same proportions hold for both types. The rounded type is used when the parts to be bolted together are nicely finished. The distance across flats W is made equal to one and one half times the diameter, plus one eighth inch, or

$$W = 1^{1}/_{2}\,d + ^{1}/_{8}''$$

The thickness of the bolt head is made equal to one half the distance across flats, or

$$T = ^{3}/_{4}\,d + ^{1}/_{16}''$$

The thickness of the nut is made equal to the diameter in all cases. These same formulae hold good for both the hexagonal and square forms. Fig. 112 shows the square form.

The radii for the various arcs are shown on the figures, and when not given in terms of the diameter are obtained from the construction, as indicated. The distance across corners is generally found by construction, as indicated in Fig. 110, by drawing a line xy at 30 degrees with the base of the head.

x-z = one half distance across flats

x-y = one half distance across corners

It should be noted that the radii R and R_1 of Fig. 111, are both drawn from the same center. The length of the radius R_1 is found by construction when drawing the bolt head or nut. When

48

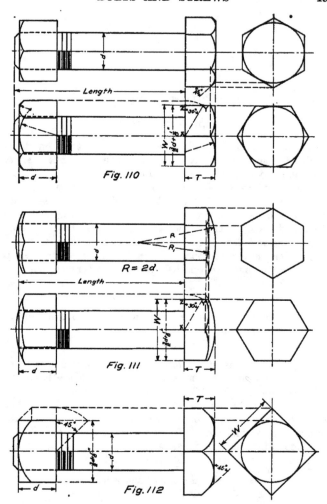

Fig. 110

R = 2d.

Fig. 111

Fig. 112

heads or nuts are finished or machined, the distance across flats is often made $1/16$ inch smaller than standard, in which case

$$W = 1^1/_2 d + ^1/_{16}''$$

The proportions of bolt heads and nuts are collected in the following list, which also gives some approximate values to be used when drawing to small scale, or where exact size is not important.

		Exact	Approximate	
Diameter of bolt.............	d	d	d	d
Distance across flats.........	W	$\frac{1}{2}d + \frac{1}{8}''$	$1\frac{3}{4}d$	
Thickness of bolt head.......	T	$\frac{3}{4}d + \frac{1}{16}'' = \frac{W}{2}$	$\frac{7}{8}d$	
(Hex.) distance across corners..	C_H	$1.155W$	$1\frac{3}{4}d + \frac{1}{8}$	$2d$
Thickness of nut.............	d	d	d	d
(Square) distance across corners	C_s	$1.414W$	$2.3d$	

Bolts. — A through bolt is one which extends through two pieces, and carries a nut, as shown in Fig. 113. Care must be taken to allow sufficient thread to insure the two pieces being held firmly together. For this reason, the distance from the end of

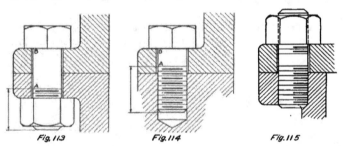

Fig. 113   Fig. 114   Fig. 115

the thread at A to the under side of the head B must be less than the thickness of the two flanges. Since the bolt head and nut are standard only three dimensions are necessary when specifying a bolt. These are, diameter, length from under side of head to end of bolt, and length of thread measured from the end of the bolt.

A tap bolt is a bolt which makes use of a part of the machine to take the place of a nut, as shown in Fig. 114. To be sure that the two pieces will be held firmly together, the distance AB must be less than the thickness of the flange.

Studs. — A stud bolt or stud is a cylindrical bar having threads on both ends (Fig. 115). Studs are used when there is not room enough for through bolts, and where there is danger of a tap bolt rusting in. Cylinder heads for steam or water machinery are familiar examples. In such cases the heads have to be taken off frequently, and if tap bolts were used, the threads might rust in, and break when an attempt to remove them was made. If successfully removed several times, the thread would be worn so as to become loose and render the keeping of a tight joint difficult. When a stud is put in place it becomes part of the casting, and the wear then comes on the nut and stud, both of which are made of wrought iron. The material will stand

Fig. 116

the wear much better than cast iron. A small amount of oil on the outer end of the stud will prevent the nut from rusting on.

Threaded Holes. — Holes for bolts and studs are generally threaded by using taps. Machinist taps come in sets of three,

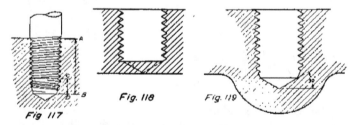

Fig. 117   Fig. 118   Fig. 119

designated as taper, plug, and bottoming taps (Fig. 116). The operation is as follows: First a hole is made with a drill having a diameter about equal to the root diameter of the screw. Such a drill is called a tap drill. The thread is then cut by inserting and turning in the taps illustrated, and in the order given. The use of the bottoming tap is often omitted, as it is seldom necessary to have threads to the very bottom of the hole (the reader is referred to catalogs of machinists' tools for further informa-

tion). Unless it is desired to have a stud jam at the bottom of
a hole, clearance, CD, should be allowed, as shown in Fig. 117.
The depth of the hole is the distance AB. If necessary the

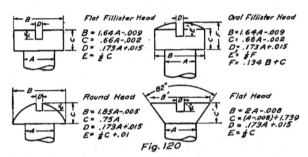

Fig. 120

thread may be carried to the bottom of the hole and even the
drill point may be ground off so that a flat bottom hole may be
obtained as in Fig. 118. This will prevent the drill from pointing
or breaking through, as indicated by the dotted lines. A better

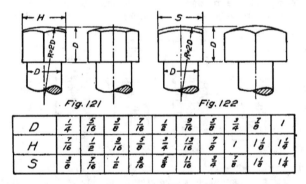

Fig. 121   Fig. 122

D	$\frac{1}{4}$	$\frac{5}{16}$	$\frac{3}{8}$	$\frac{7}{16}$	$\frac{1}{2}$	$\frac{9}{16}$	$\frac{5}{8}$	$\frac{3}{4}$	$\frac{7}{8}$	1
H	$\frac{7}{16}$	$\frac{1}{2}$	$\frac{9}{16}$	$\frac{5}{8}$	$\frac{3}{4}$	$\frac{13}{16}$	$\frac{7}{8}$	1	$1\frac{1}{8}$	$1\frac{1}{4}$
S	$\frac{3}{8}$	$\frac{7}{16}$	$\frac{1}{2}$	$\frac{9}{16}$	$\frac{5}{8}$	$\frac{11}{16}$	$\frac{3}{4}$	$\frac{7}{8}$	$1\frac{1}{8}$	$1\frac{1}{4}$

method is to put a boss on the casting opposite the hole, and
then use a regular drill and plug tap (Fig. 119).

Machine Screws. — Small screws are made with a variety of
forms of heads. They are especially adapted for use with small
parts of machines. Fig. 120 shows the various forms of heads,
and the proportions as recommended by the American Society
of Mechanical Engineers. The sizes of machine screws are

designated by numbers. Diameters range from .060 inches to
.450 inches.

Cap Screws. — For many purposes bolts having different
dimensions from the U. S. Standard are desirable. Hexagonal
and square cap screws are shown in Figs. 121 and 122. The
distance across flats is less than the U. S. Standard, and the
thickness is greater. Cap screws are also made with heads similar
to those shown for machine screws. Cap screws are designated
by their diameter in inches. The diameters are in even fractions
of an inch, starting at $1/4''$.

Cap Nuts. — Where an especially finished appearance is desired, cap nuts may
be used to conceal
the ends of studs.
They are frequently
seen on polished cyl-

inder heads, and similar places. Several forms of cap nuts are
shown in Figs. 123, 124, and 125.

Set Screws. — For holding pulleys on shafts, and otherwise
preventing relative motion, set screws may be used. Several
forms are illustrated. Any combination of point and head may

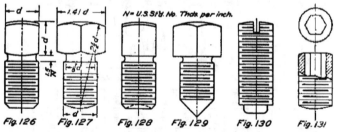

Fig. 126   Fig. 127   Fig. 128   Fig. 129   Fig. 130   Fig. 131

be obtained. Some proportions are shown in Figs. 126 to 131.
A projecting set screw on a revolving pulley is a source of great
danger, and should be avoided. The many forms of headless
and hollow set screws on the market render the use of other
forms unnecessary in such cases.

The relative holding power of the different forms of ends of
set screw are given by Professor Lanza in the A. S. M. E. "Transactions," Volume 10. Average results of tests on four kinds are
as follows:

A. Flat end, $^9/_{16}$ inch diameter, 2064 pounds
B. End rounded, $^1/_2$ inch radius, 2912 pounds
C. End rounded, $^1/_4$ inch radius, 2573 pounds
D. Cup shaped end, 2470 pounds

The set screws were all $^5/_8$ inches in diameter, and were tightened with a pull of 75 pounds on a 12 inch wrench.

Locking Devices. — The vibration of machinery often causes nuts to become loose if they are not provided with some form of locking device. The commonest method is to use two nuts. They may be full size, or one of the arrangements shown in Fig.

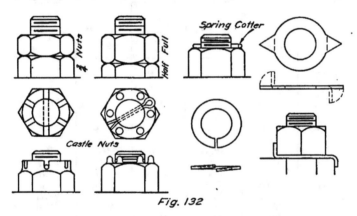

Fig. 132

132. The castle nut illustrated forms a good method. Lock washers consisting of a piece of sheet metal are effective. One corner is turned down, and another corner is turned up, as illustrated.

The following table gives the dimensions for U. S. Standard bolt heads and nuts.

DIMENSIONS OF U. S. STANDARD BOLT HEADS AND NUTS

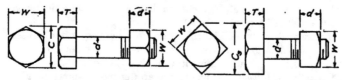

d Diameter of Bolt	W Flats or Short Diameter	C Corners or Long Diameter	d Thickness of Nut	T Thickness of Bolt Head	C_s Corners or Long Diameter
$1/4$	$1/2$	$27/64$	$1/4$	$1/4$	$23/32$
$5/16$	$19/32$	$11/16$	$5/16$	$19/64$	$27/32$
$3/8$	$11/16$	$51/64$	$3/8$	$11/32$	$31/32$
$7/16$	$25/32$	$29/32$	$7/16$	$25/64$	$1\,7/64$
$1/2$	$7/8$	$1\,1/64$	$1/2$	$7/16$	$1\,1/4$
$9/16$	$31/32$	$1\,1/8$	$9/16$	$31/64$	$1\,3/8$
$5/8$	$1\,1/16$	$1\,15/64$	$5/8$	$17/32$	$1\,1/2$
$3/4$	$1\,1/4$	$1\,29/64$	$3/4$	$5/8$	$1\,3/4$
$7/8$	$1\,7/16$	$1\,43/64$	$7/8$	$23/32$	$2\,1/32$
1	$1\,5/8$	$1\,7/8$	1	$13/16$	$2\,6/8$
$1\,1/8$	$1\,13/16$	$2\,3/32$	$1\,1/8$	$29/32$	$2\,9/16$
$1\,1/4$	2	$2\,9/16$	$1\,1/4$	1	$2\,43/64$
$1\,3/8$	$2\,3/16$	$2\,17/32$	$1\,3/8$	$1\,3/32$	$3\,3/32$
$1\,1/2$	$2\,3/8$	$2\,3/4$	$1\,1/2$	$1\,3/16$	$3\,23/64$
$1\,5/8$	$2\,9/16$	$2\,15/16$	$1\,5/8$	$1\,9/32$	$3\,5/8$
$1\,3/4$	$2\,3/4$	$3\,5/16$	$1\,3/4$	$1\,3/8$	$3\,57/64$
$1\,7/8$	$2\,15/16$	$3\,11/32$	$1\,7/8$	$1\,15/32$	$4\,5/16$
2	$3\,1/8$	$3\,5/8$	2	$1\,9/16$	$4\,27/64$

DEPTH OF TAPPED HOLES AND DISTANCE FOR SCREW TO ENTER

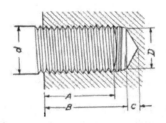

d Diameter of Screw	D Diameter of Tap Drill	B Depth of Hole	C Allowance for Drill Point	A Distance for Screw to Enter
$1/4$	$13/64$	$7/16$	$1/16$	$3/8$
$5/16$	$17/32$	$9/16$	$5/64$	$7/16$
$3/8$	$5/16$	$11/16$	$3/32$	$9/16$
$7/16$	$3/8$	$3/4$	$7/64$	$5/8$
$1/2$	$27/64$	$13/16$	$1/8$	$11/16$
$9/16$	$31/64$	$15/16$	$9/64$	$13/16$
$5/8$	$17/32$	1	$5/32$	$7/8$
$3/4$	$41/64$	$1 1/4$	$3/16$	1
$7/8$	$3/4$	$1 1/2$	$7/32$	$1 1/4$
1	$55/64$	$1 5/8$	$1/4$	$1 3/8$
$1 1/8$	$61/64$	$1 3/4$	$9/32$	$1 1/2$
$1 1/4$	$1 5/64$	2	$5/16$	$1 3/4$
$1 3/8$	$1 11/64$	$2 1/4$	$11/32$	$1 7/8$
$1 1/2$	$1 19/64$	$2 1/2$	$3/8$	$2 1/8$
$1 5/8$	$1 13/32$	$2 5/8$	$13/32$	$2 1/4$
$1 3/4$	$1 1/2$	$2 3/4$	$7/16$	$2 3/8$
$1 7/8$	$1 5/8$	3	$15/32$	$2 5/8$
2	$1 23/32$	$3 1/8$	$1/2$	$2 3/4$

CHAPTER VIII

RIVETING

Riveting. — Since machines and structures cannot be made in one piece some means of fastening the parts together must be used. For many purposes where a permanent fastening is required, rivets are used. A rivet is a bar of metal having a head made on one end and a length sufficient to allow forming a head on the other end after being put into place. The holes for rivets may be either punched or drilled. As punching injures the metal, drilled holes are better for boiler or other pressure work.

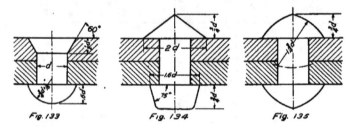

Fig. 133   Fig. 134   Fig. 135

Holes are made $1/16$ inch larger diameter than the rivets used in them. Thus a one-inch rivet is $15/16$ inch diameter before driving.

The computations for pitch and efficiency of joints, matters relating to design, are beyond the scope of this work, but the following articles will suffice for drawing purposes.

Rivet Heads. — The forms of rivet heads are shown in Figs. 133, 134, and 135. The countersunk head and the button head are illustrated in Fig. 133. These forms are used for structural work. For pressure work the cone head or pan head of Fig. 134 may be used, or the common form of Fig. 135.

Lap Joints. — When two plates lap over each other and are held by a row of rivets as in Fig. 136 it is called a single riveted lap joint. A double riveted lap joint is shown in Fig. 137. The distance between the centers of two rivets in the same row is called the pitch. The distance from the center line of the rivets to the edge of the plate is called the lap. The lap is commonly made equal to one and one half times the diameter of the rivet.

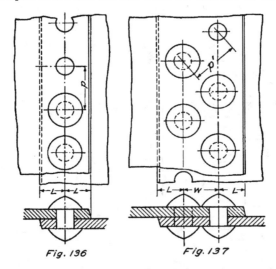

Fig. 136   Fig. 137

The distance from the center of a rivet in one line to the center of a rivet in the next line is called the diagonal pitch and may be found from the formula:

$$P' = {}^2/{}_3P + \frac{d}{3}$$

Either chain riveting (Fig. 138) or staggered riveting (Fig. 139) may be used when there are several rows of rivets.

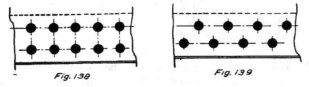

Fig. 138   Fig. 139

Butt Joints. — Three forms of butt joints are shown in Figs. 140, 141, and 142. In Fig. 140 a single butt-strap having a thickness of about one and one fourth times the thickness of the plates

may be used. Figs. 141 and 142 show single and double riveted
butt joints with two butt-straps. In such cases the butt-straps
may be $^1/_{16}$ inch thinner than the plates.

When three plates come together they must be arranged so

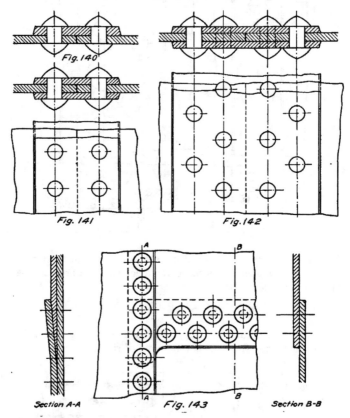

Fig. 140

Fig. 141   Fig. 142

Section A-A   Fig. 143   Section B-B

as to maintain a tight joint. One method used is shown in
Fig. 143. In order to obtain a fit one of the plates must be
thinned out.

Calking. — For many purposes rivets must make a leak tight
joint as well as hold the plates together. To assist in this a

blunt chisel is used to force or pound the edge of the plate down. This is called calking and makes a water or steam tight joint

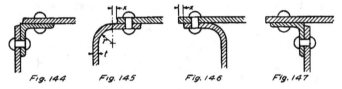

Fig. 144   Fig. 145   Fig. 146   Fig. 147

between the plates. The bevel of about 75° shown is to make the calking easier.

Miscellaneous Connections. — Some miscellaneous connections are shown in Figs. 144 to 147. Angles may be used as in Figs.

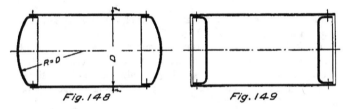

Fig. 148   Fig. 149

144 and 147 or one of the plates may be bent as in Figs. 145 and 146. In this case the radius of curvature (r) may be about two and one half times the thickness of the plate. Also note that a short straight part (x) should be provided to allow easy calking

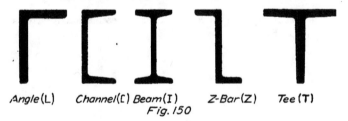

Angle(L)   Channel(C) Beam(I)   Z-Bar(Z)   Tee (T)

Fig. 150

(Fig. 145). When drawing to a small scale thin sections are sometimes blacked in as shown in Figs. 148 and 149, which also illustrate methods of closing the ends of cylindrical tanks. With rounded ends the radius of curvature may be taken equal to the diameter of the tank.

Rolled Steel Shapes. — For many constructions, rolled steel shapes are used. The dimensions and weights as well as other properties can best be obtained from the handbooks issued by

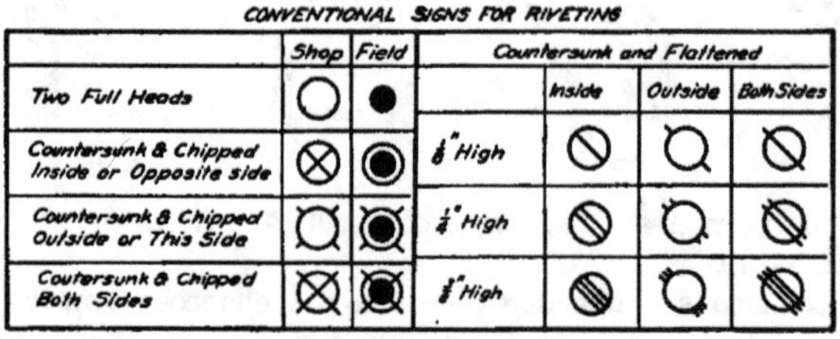

CONVENTIONAL SIGNS FOR RIVETING

	Shop	Field	Countersunk and Flattened			
				Inside	Outside	Both Sides
Two Full Heads	○	●				
Countersunk & Chipped Inside or Opposite side	⊗	◉	⅛″ High	⊘	○	⊘
Countersunk & Chipped Outside or This Side	⊘	◉	¼″ High	⊘	○	⊘
Countersunk & Chipped Both Sides	⊗	◉	⅜″ High	⊘	○	⊘

Fig. 151

the steel companies. The names of a few of the common sections are given in connection with Fig. 150.

The pitch of rivets for structural purposes may be taken at from three to six inches. The distance from the center of the rivet to the edge of the plate should generally be about two times the rivet diameter. The pitch for various sizes of rivets may be taken from the table given below.

MINIMUM RIVET SPACING

Diameter of Rivet	$\frac{1}{4}$	$\frac{3}{8}$	$\frac{1}{2}$	$\frac{5}{8}$	$\frac{3}{4}$	$\frac{7}{8}$	1
Pitch	$\frac{3}{4}$	$1\frac{1}{8}$	$1\frac{1}{2}$	$1\frac{7}{8}$	$2\frac{1}{4}$	$2\frac{5}{8}$	3

The Osborn system of conventional representation for rivets is shown in Fig. 151.

CHAPTER IX

Classes of Drawings. — The origin of a drawing is of interest, and a knowledge of how drawings are produced is essential. Roughly drawings may be divided into two classes; detail drawings and assembly drawings. These names are sufficiently descriptive in a general way. Drawings are sometimes made from a machine or part by measuring and sketching. The usual source of a detail drawing is the designer's board. Here the whole machine is laid out to scale in a more or less complete manner, the relation of one part to another is shown, and such fixed dimensions as are necessary are determined. The shapes of the various parts as required for strength and motion are worked out and drawn. From such drawings the detail draftsman works and finishes the drawings of the separate parts.

A detail drawing shows each piece separately and completely defines it (Fig. 152). The number of views is determined by what is necessary to show the shape and size of the object. A pin, shaft, or bolt can generally be shown in one view, while a casting may require two, three, or more views together with sectional and auxiliary views. The main views should always be arranged in strict conformity to the rules of projection. The third quadrant is used exclusively for this purpose. Auxiliary views and sections may be placed in other positions but explanatory notes should always be used to define them as explained in Chapter X.

The size of paper and the scales to use have been treated in other chapters. Use a scale that will show the object clearly and that will not require crowding of the dimensions. In general it is better not to use more than one scale on the same sheet. To this end large and small pieces would not be put on the same sheet. There are many concerns where each part is drawn on a sheet by itself. The character of the work will determine the practice in this respect.

62

It is generally well to draw large castings separately and to group small parts together as:

Small Castings,
Bronze and Composition Castings,
Forgings,
Bolts and Screws.

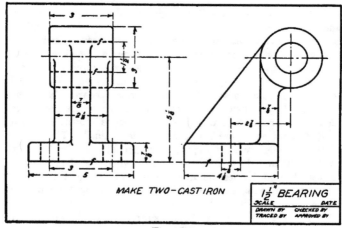

MAKE TWO—CAST IRON

$1\frac{1}{2}"$ BEARING

SCALE	DATE
DRAWN BY	CHECKED BY
TRACED BY	APPROVED BY

Fig. 152

Special Detail Drawings. — Special detail drawings are sometimes made for the different classes of workmen. These might be classed as follows:

Pattern Drawings,
Forging Drawings,
Machinist's Drawings,
Stock Drawings.

There are many advantages to this system where a large number of parts are made as each workman is given only such information as concerns him. As pattern dimensions are used only when the pattern is made or for alterations they complicate the drawing and can better be left off the machinist's drawing. One method is to put the pattern dimensions and information on the tracing in pencil, make several blueprints, and erase the pencil

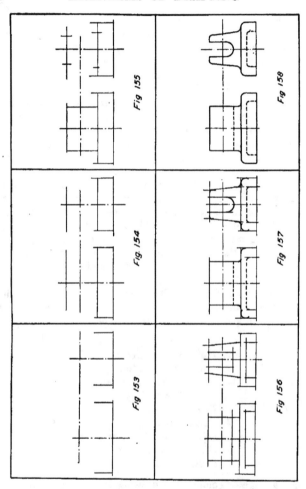

information from the tracing. Gasolene applied with a soft cloth is excellent for this purpose. For forgings two separate drawings will be necessary, one for the blacksmith and one for the machinist. The saving in time will make up for the expense of the extra drawing in most cases.

How to make a Drawing. — A detail drawing is started by first locating the main center lines as shown in Fig. 153 for the necessary views. Next "block in" the fixed dimensions in all views and from them work out the shape of the object. The small circles, fillets, etc. should be drawn last. Figs. 153 to 158 show the drawing for a slide valve in the various stages of making.

After completing the drawing in pencil it is ready to be inked on paper or traced.

Tracing. — Most drawings are now inked on tracing cloth. This is a translucent linen cloth. There are many grades, some nearly transparent. One side of the cloth is generally shiny or glazed and the other dull. Either side may be used but the dull side is to be preferred. The cloth is tacked down over the pencil drawing and the lines inked in as though they were on the cloth. The surface of the cloth should be rubbed over with powdered chalk and then the chalk thoroughly removed. A clean blotter will serve the same purpose. The fine thread running at the edges of the cloth should be torn off before using to prevent wrinkling. As the cloth is absorbent it should be protected from moisture.

Order for inking Lines. — The weight of line to be used has been discussed in the first chapter. First ink the center lines using a fine dot and dash line. The order of inking then is:

1. *Small circular arcs and circles.*
2. *Large circular arcs and circles.*
3. *Irregular curved lines.*
4. *Straight horizontal lines.*
5. *Straight vertical lines.*
6. *Dotted circular arcs.*
7. *Dotted lines.*
8. *Witness and dimension lines.*
9. *Dimensions, notes, title.*
10. *Section lining.*

When a large or complicated drawing is to be inked it is advisable to ink one view at a time or to start only so much as can be completed on the same day. If a view is left uncompleted it will generally be found very difficult to join the various lines, because the cloth is very sensitive to atmospheric changes which cause it to stretch.

Assembly Drawings. — An assembly drawing shows the parts of a machine in their proper relation to one another. There are many kinds of assembly drawings, some of which will be described.

An *Outline* or *Setting* drawing is frequently made to show the appearance of the machine, give center distances, and overall dimensions. Such drawings are used to illustrate the machine to prospective customers, to lay out the foundation, and for locat-

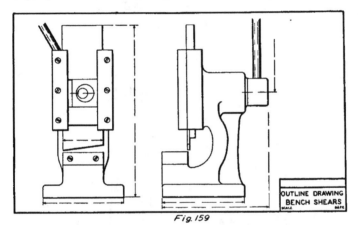

OUTLINE DRAWING
BENCH SHEARS

Fig. 159

ing the machine in its building. Fig. 159 shows one form of such a drawing.

An *Assembly Working Drawing* is often made when only a few of the machines are to be constructed. Such a drawing might contain a number of part views or sections. It would be completely dimensioned so that no separate or detail drawings would be required. Fig. 175 shows such a drawing.

Part Assembly Drawings are sometimes made giving a few pieces in their proper relation to each other and either partly or completely dimensioned. When completely dimensioned no further detail drawings are made.

Assembly drawings made to show the sizes, location, and method of fastening pipes and wires are called piping or wiring diagrams or drawings, depending upon how completely they are figured.

Erection Drawings are an important class of assembly drawings. They show the proper order of putting the parts together, dimen-

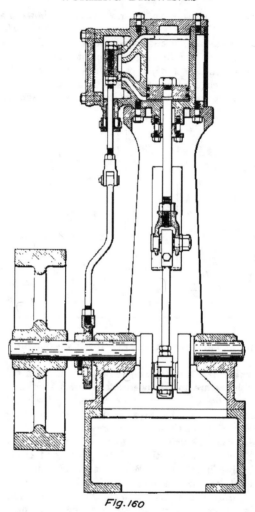

Fig. 160

sions, such as center distances, which must be exact, give the
location of oil holes, valves, switches, etc., and methods of making
adjustments.

Diagram Drawings are used by many concerns. These comprise a sectional or external view of the whole of the machine upon which the parts can be numbered or named. Such a drawing frequently contains a list of the parts, drawing numbers, pattern numbers, materials, weight, and other information.

Outline drawings are often used for catalogs, advertising, and similar purposes. Some of the points to be considered are given

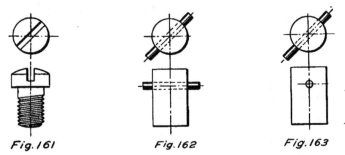

Fig. 161   Fig. 162   Fig. 163

in the following list. The one upon which emphasis must be put will depend upon the use to which the drawing is to be put.

1. Get the important points.
2. Sense of proportion.
3. Suggestion.
4. Simplicity (few lines).
5. Record peculiarities in shape or design.
6. Use notes if necessary.
7. Number of machine.
8. Name of manufacturer.
9. Trade names.
10. Use of shading.
11. Not necessarily to scale.

Show Drawings are sometimes made. These are often in the nature of a picture in which the center lines and dimensions are left off (Fig. 160). Line shading as explained in a later chapter is often used. A good effect may sometimes be obtained by mass shading with a soft pencil, using the dull side of the tracing cloth. For more particular work on paper, india ink tinting applied with a brush can be used.

Exceptions to True Projection. — There are many cases where true projection is departed from in the interests of simplicity and clearness. Figs. 161, 162, and 163 show a few cases. The slot in the screw is drawn at 45° in the top view but is not projected

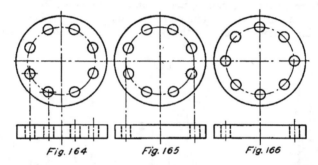

Fig. 164   Fig. 165   Fig. 166

to the elevation. The same practice is followed for holes and pins. The location of bolt holes is another illustration. The front view of Fig. 164 shows the true projection of the bolt holes. The front view of Fig. 165 shows the preferable method which

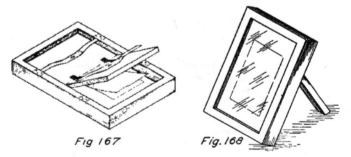

Fig 167   Fig. 168

locates the centers of the bolt holes at a distance apart equal to the diameter of the circle of drilling. In such cases the other holes need not be projected as they add nothing to the information conveyed by the drawing. When holes are drilled as in Fig. 165 they are said to be "Two Up" or off centers, and when located as in Fig. 166 they are said to be "One Up" or on centers. Pipe flanges on elbows and fittings are usually drilled "Two Up" and

with four, eight, or some multiple of four holes, so that the flanges
can be turned at right angles.

Other exceptions to true projection are discussed in the chapter
on sections.

Blueprints. — The object of making tracings is to provide a
convenient means for obtaining several copies of the original
drawing. The most common method is by the blueprinting
process. Blueprint paper is paper which has been coated with
iron salts which are sensitive to light. The method of making
blueprints is as follows:

Place the tracing with the right side or inked side next to the
glass of a printing frame as shown in Fig. 167. Next place a
piece of blueprint paper on the tracing with the coated side down.
Follow this with the felt pad and close the frame. Expose to
the direct sunlight as indicated in Fig. 168. The length of the
exposure varies from 30 seconds in strong sunlight with rapid
printing paper to three or four minutes under the same conditions
with slow printing paper. The time can best be found by trial,
as the age of the paper and the brightness of the light all exert
an influence. After exposing, the paper should be removed and
thoroughly washed. The excess water may be blotted off and
the print hung up to dry. New paper has a yellow color on the
coated side. After exposure this changes to a gray-bronze except
where the lines of the tracing prevent the light from reaching it.

Electric light is very generally used in the larger mechanical
factories for making blueprints. Machines for this purpose as
well as many other methods of duplication are described in drawing supply catalogs to which the reader is referred.

CHAPTER X

SECTIONS

Sectional Views. — Probably the most useful form of conventional representation is the sectional view obtained by an

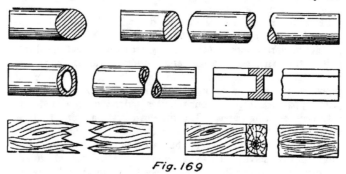

Fig. 169

imaginary cutting plane described in Chapter IV. Free use of sections often saves much time as well as possibility of mistakes in reading drawings for constructions which have complicated

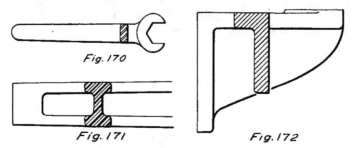

Fig. 170

Fig. 171

Fig. 172

cores. The choice of views should be made with care and for a definite purpose, never for appearances. There are many special sections, some of which are described in this chapter. An article

by the author, "Sections of Ribs and Symmetrical Parts," in
"Machinery," June, 1915, gives further applications.

Broken and Revolved Sections. — When a long piece of uniform cross section is to be represented, a larger scale can be used

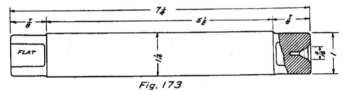

Fig. 173

by "breaking" the piece. The manner of breaking generally
indicates the form of cross section and material as in Fig. 169.
The break is made free hand but should be carefully done. The
two sides should appear to match, that is, if the sectioning comes
on the upper side of one half it should come on the lower side of

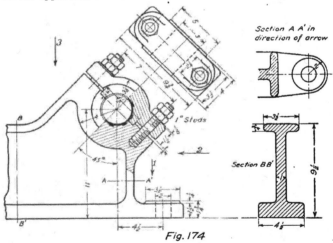

Fig. 174

the other. A similar method of "set in" sections is often used
for such conditions as are present with wrench handles, pulley
arms, brackets, hand wheels, and rods. Figs. 170 to 173 show
some examples.

Location of Sectional Views. — When conditions permit, sectional views should be placed according to the laws of projection

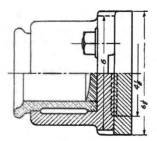

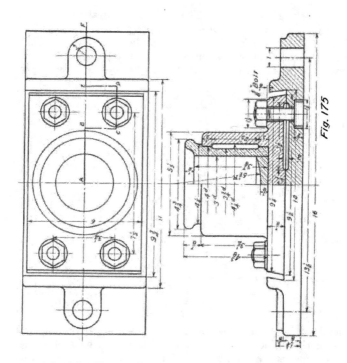

Fig. 175

as explained in Chapter IV, and are drawn in the same manner as the other views by assuming a part of the machine or parts to have been removed. When many sections are required or other reasons make it necessary to place the sectional views in another location, arrows and notes should be used to explain them as shown in Fig. 174. Extra sectional views are often very useful in explaining parts of a machine or details of a part.

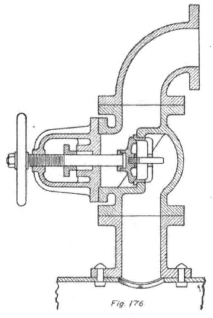

Fig. 176

Since the cutting plane is imaginary it need not be continuous; thus several sections may be used and the views represented as though occurring on a single plane. This is illustrated in Fig. 175, where the cutting plane is changed as shown in the top view. Thus the front section is taken on the plane A, B, C, D, E, F, and the side section on a plane through the center.

Objects not Sectioned. — When a full view will serve the same purpose just as well a sectional view should not be used. This is true in the case of shafts, bolts, nuts, screws, rivets, keys, pulley

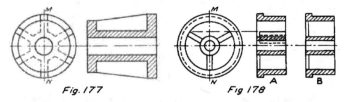

Fig. 177   Fig 178

arms, etc., which are very seldom drawn in section except when the cutting plane is at right angles to the long dimension. This treatment of a section is shown in Fig. 176.

Dotted Lines on Sectional Views. — Very often a sectional view contains only the outline of the sectioned surfaces and the full lines which appear. How much of the part behind the plane

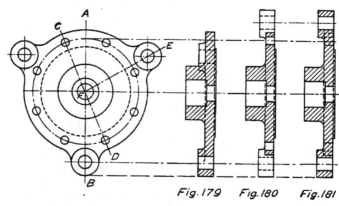

Fig. 179   Fig. 180   Fig. 181

of the section should be represented must be determined for each particular case. When an object is represented by a view made up of one half in section and one half exterior most or all of the

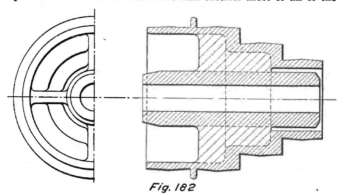

Fig. 182

dotted lines may be omitted from both halves, as was done in Fig. 175.

Sections of Ribs and Symmetrical Parts. — Ribs, arms, and gear teeth are not ordinarily sectioned even though they appear

on the plane of the section. Figs. 177 and 178 illustrate such
cases. In Fig. 177 the plane MN passes through the ribs, but is
not sectioned in the other view as it would give a false impression
of solidity. In Fig. 178 the true projection without sectioning
the rib is shown at A, while the usual conventional section is
shown at B.

The representation of a cylinder head in Figs. 179, 180, and 181,
shows a similar case. A true section on the plane AB is given
in Fig. 179. In Fig. 180 the section is taken on CD and revolved
into the position of AB. The bolt holes and lugs are then located
at their true distances from the center. By this means one view
could be made to represent the cylinder head by adding a note
to give the number of lugs. An alternate method is shown in
Fig. 181, where the section FE is revolved. The idea in all cases
is to avoid a view which might in any way be confusing and to
convey the true shape clearly.

When a rib occurs on the plane of a section and it is necessary
to distinguish it, coarse sectioning may be employed as in the
cone pulley of Fig. 182 where the ribs are sectioned but alternate
lines are omitted. A note giving the number and thickness of
the ribs would allow the end view to be dispensed with. Observe
that the half end view is bounded by the center line and not by
a full line, as the pulley has not been actually cut in half.

CHAPTER XI

DIMENSIONING

Purpose of Dimensions. — The purpose of dimensions is to give the necessary figures for constructing machine parts and putting them together. A drawing gives the shape of an object,

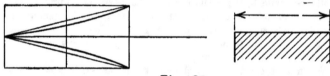

Fig. 183

the dimensions tell the size. These are two operations and both should be kept in mind.

Dimension Lines. — Dimension lines show where the figures apply to the drawing. They are terminated by arrow heads. The arrow heads should be about twice as long as they are wide.

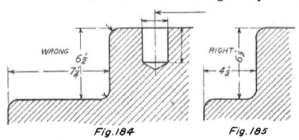

Fig. 184   Fig. 185

Fig. 183 shows the construction of an enlarged arrow head, and its proportions. Fine full red ink lines are sometimes used for dimension, center, and witness lines. The arrow heads and figures are always black. The figures and notes should always be placed so as to read from the lower or right hand side of the

77

drawing. Never use slant fraction lines. In most cases it is considered bad practice to place the figures upright as shown in Fig. 184 where the figures may be easily read with the wrong dimension lines. Fig. 185 shows a better arrangement. The witness and dimension lines should be as fine as possible so as not to conflict with the lines of the drawing. In the interest of clearness there should be as few lines as possible crossing each

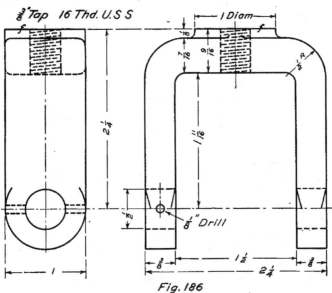

Fig. 186

other. The center lines and object lines have only one purpose and should never be used as dimension lines. Generally the dimension lines can be kept outside of the views, thus keeping the size and shape of the object separate. In such cases place the larger dimensions outside the smaller ones as in Figs. 186 and 188. Fig. 187 shows a poorly dimensioned drawing of a pump plunger and Fig. 188 the same piece properly dimensioned. Finished surfaces are ordinarily indicated by a letter "f" placed across the line which represents the surface to be machined.

Elements of Dimensioning. — Constructions can be separated into parts and these parts can then be divided into geometrical

solids. Each of the solids can then be dimensioned and their
relation to each other fixed. Figs. 189 to 193 show a prism, a
pyramid, a cone, and a cylinder with dimensions. Figs. 194,
195, and 196 show combinations. Note that the same location
of dimensions is maintained. In dimensioning cylinders give the
diameters on the elevation as in Fig. 195. Placing of the five

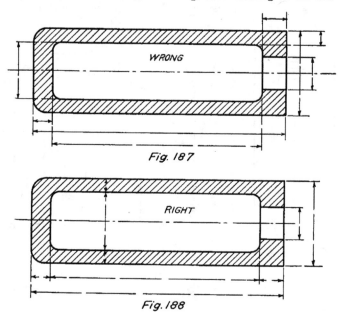

Fig. 187

Fig. 188

diameters on the end view would result in crowding as well as
inconvenience in reading figures placed at an angle. Always
give a diameter in preference to a radius if the part is a complete
cylinder. For quarter rounds, fillets, and part circles give the
radius.

General Rules. — To dimension a drawing successfully the
construction of the pattern, machining, fitting, and putting together of the machine must be gone over. It is necessary to keep
constantly in mind the person to whom the drawing is addressed
and the purpose for which it is to be used.

Hints:

Do not hurry,
Give sizes of pieces for the pattern maker,
Give sizes and finish for the machinist,
Give assembly dimensions,
Give office dimensions,
Give notes where needed.

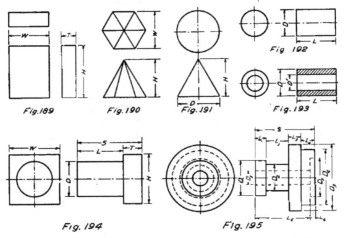

Fig.189   Fig.190   Fig.191   Fig.192   Fig.193

Fig. 194   Fig. 195

It is necessary to remember that surfaces and not lines are being located. The dimensions of the piece must be kept in mind. Detail drawings are generally made to serve both pattern maker and machinist, and the figures indicate the size of the finished piece. The pattern maker is left to make required allowances for finish, shrink, and draft. In the case of forgings two drawings are sometimes made, one for the blacksmith giving the rough sizes, and another for the machinist giving the finished sizes.

Systems of Dimensioning. — Four general systems of dimensioning may be mentioned as follows:

1. *All figures outside of the object lines.*
2. *All figures inside of the object lines.*
3. *All figures given from two reference lines at right angles to each other.*
4. *A combination of the preceding three systems.*

The four systems are illustrated in Figs. 197 to 200. The first method is to be favored as the dimension lines and figures are kept separate from the interior and allow details to be easily seen. The size and shape are separated. The second method may be used when there is little detail within the view. It preserves the outline of the view but often there is confusion due to the crossing of the lines and crowding of the figures. The third method is particularly adapted to plate work and laying out where holes must be carefully located.

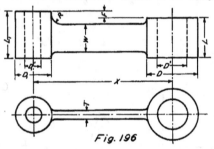

Fig. 196

The fourth method is the one generally used but making it

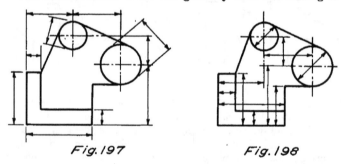

Fig. 197   Fig. 198

conform to the first system by placing dimensions outside whenever it is conveniently possible.

Location of Dimensions. — Facility in manufacture should be a motto in dimensioning. The figures must be so placed as to be easily found and perfectly clear in their meaning when found. Select that view which most completely defines the object and start with it first. If an assembly drawing, dimension only one piece at a time and finish all views of that one piece before starting another. Put on similar dimensions at the same time, as diameters, lengths, etc. Do not jump from one piece to another. Work from the more important dimensions to those of less importance.

See that all center distances are given. Consider the effect of location upon ease of reading the drawing. Similar pieces should be dimensioned in exactly the same way. Fig. 201 shows a gland,

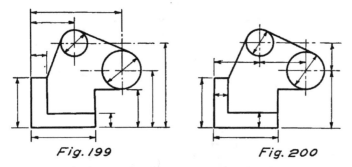

Fig. 199   Fig. 200

Fig. 202 a pump valve, and Fig. 203 a cylinder head. They are all similar pieces and the dimensions are located in the same places on each. In the three figures the similar dimensions are indicated by the letters A, B, C, etc.

By observing such methods a system of dimensioning can be

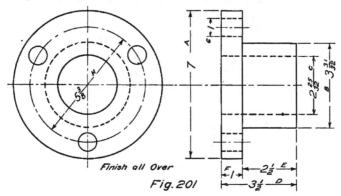

Finish all Over

Fig. 201

employed which will save a great deal of time and many mistakes and omissions. It is seldom necessary to repeat the same dimension on a drawing. Drilling is generally best located in the view where it shows in plan, that is, in the view where it is laid out. Diameters are always clearer when shown on a section or elevation rather than on an end view. The drilling for flanges is dimensioned by giving the diameter of the bolt circle and the size of bolt holes or bolts. The holes are understood to be equally spaced unless noted otherwise.

Shafting. — Shafting should be dimensioned by giving the diameters and lengths together with the sizes of keyways and pins and their location. Shafting is made from various grades of wrought iron and steel. For many purposes cold rolled shafting is generally used. This is shafting which has been cleaned of scale and rolled under pressure. It can be used without the

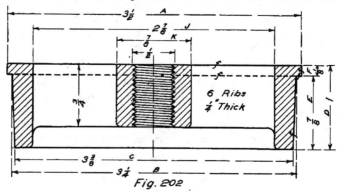

Fig. 202

necessity for turning and is considerably strengthened by the surface skin which comes from the rolling process. Hot rolled shafting is black and must be turned to size before using. Usual sizes are:

NOMINAL DIAMETERS OF SHAFTING

$1^1/_4$	$2^1/_2$	4
$1^1/_2$	$2^3/_4$	$4^1/_2$
$1^3/_4$	3	5
2	$3^1/_4$	$5^1/_2$
$2^1/_4$	$3^1/_2$	6

These are nominal sizes and are $^1/_{16}$ inch larger than actual diameter. Thus a 2-inch shaft is $1^{15}/_{16}''$ actual diameter. Common lengths vary up to 24 feet. Special shafts have to be forged of steel suitable for the particular purpose. A shaft drawing is shown in Fig. 204 with the positions of the dimensions.

Tapers. — Various methods are in use for designating tapers. Figs. 205, 206, and 207 show ways of indicating the two diameters and the length. Sometimes a note is employed giving the taper

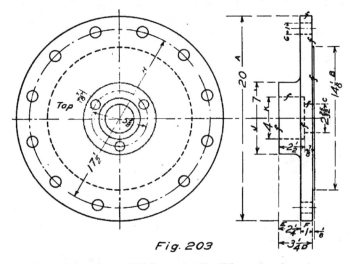

Fig. 203

per foot of length as, "$^3/_4$" *per foot.*" When the slope is considerable it may be given as 1:1, indicating a 45° slope. In other cases, the angle may be given in degrees. In addition there are

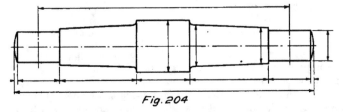

Fig. 204

a number of standard tapers in use such as B & S (Brown & Sharpe), Morse Tapers, Reed Lathe Center Tapers, Jarno Tapers, and Sellers Tapers. In such cases the taper is indicated by a number which fixes the three dimensions, large diameter, small diameter, and length. A machinist's handbook should be consulted for complete information.

Small Parts. — There are many small parts such as shafts, pulleys, etc., which can be defined in one view by using a note to give the missing dimensions. When clearness is not sacrificed it is this method in many cases. Small details which are standardized do not need to be completely dimensioned. This is true for *bolts and screws, standard tapers, piping, wire, sheet metal, rope, chain, pins, rolled steel shapes.*

Methods of Finishing. — In connection with dimensions the limits of accuracy for all fits should be given. The method of

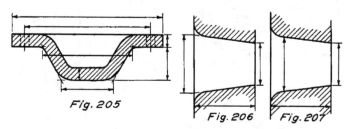

Fig. 205   Fig. 206   Fig. 207

finishing is given in another chapter, and should be indicated by a note and arrow.

1. *Rough.*	8. *Loose fit.*
2. *Rough turned.*	9. *Driving fit.*
3. *Ground.*	10. *Scraped.*
4. *Polished.*	11. *Finished.*
5. *Reamed.*	12. *Drilled.*
6. *Cored.*	13. *Chipped.*
7. *Running fit.*	14. *Spot faced.*

Checking Drawings. — The checking of a drawing is one of the important duties of most draftsmen. Whenever possible it should be done by someone who has not worked on the drawing. The first thing to do is to see if the drawing can be used without unnecessary difficulty, and to see if the parts are such as will fit and operate successfully. There must be clearance for moving parts. Then observe if sufficient views are given to completely determine the parts, and that all dimensions necessary for machining and erecting are given and that they are properly located. Check the correctness of all figures by use of the scale and by computation. All notes should contain a clear statement and be carefully located. Standard parts should be used where possible. See that the fewest number of different sizes of bolts and similar small parts are used. Consider the materials of which the parts are made, the construction of the patterns and cores, and the method of machining. A valuable article on "How Machinery Materials and Supplies are Sized" is given in "Machinery," February, 1916.

CHAPTER XII

MACHINE CONSTRUCTION

Machine Operations. — The parts of machines which come from the foundry, forge, or rolling mill generally require finishing, such as machining to size, drilling, tapping of holes, etc., before they can be assembled in the machine of which they are to be a part. A knowledge of what is involved in the processes of machining is important to the machine draftsman. The principal machine operations are turning, drilling, boring, planing, and milling. The machines used are lathes, drills, boring mills, planers, milling machines, shapers, etc.

In order to pursue the subject of drawing with profit at least

Fig. 208   Fig. 209

one book on machine tools should be purchased and studied. The advertising pages as well as the reading pages of such magazines as " American Machinist " and "Machinery" are further sources of information which should not be neglected. Every opportunity should be availed of to observe and study work as it is carried out in pattern shop, forge, foundry, and machine shop. Such knowledge is invaluable and will often enable the draftsman materially to reduce the expense of production by simplifying or adapting his designs.

Drills. — Drills are used for making holes of comparatively small diameter. Two forms of drills are shown in Figs. 208 and 209. The first is a flat drill and the second a twist drill. The latter is the form in general use. Drills are used in different forms of machines. Look up the following in the advertising pages of "American Machinist" or "Machinery": Sensitive Drill, Drill Press, Multiple Drill.

The Steam Engine. — It is important for the draftsman to learn the names of the parts of the steam engine. Fig. 210 shows the principal parts.

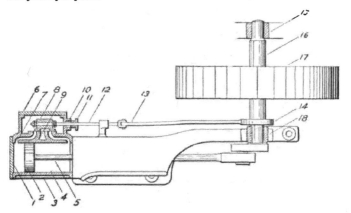

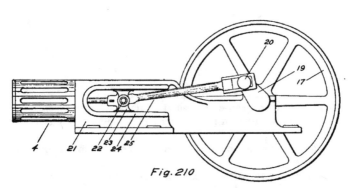

Fig. 210

1. Cylinder head.	8. Slide valve.
2. Piston.	9. Exhaust port.
3. Casing or lagging strip.	10. Valve rod stuffing box.
4. Cylinder.	11. Valve rod gland.
5. Piston rod.	12. Valve rod.
6. Steam chest cover.	13. Eccentric rod.
7. Steam port.	14. Eccentric.

15. Outer bearing.
16. Main shaft.
17. Fly wheel.
18. Inner bearing.
19. Crank.
20. Crank pin.

21. Frame.
22. Crosshead pin.
23. Crosshead.
24. Crosshead guide.
25. Connecting rod.

Steam is admitted to alternate sides of the piston by means of the slide valve which is actuated by the eccentric through the eccentric rod. The piston transmits the pressure of the steam

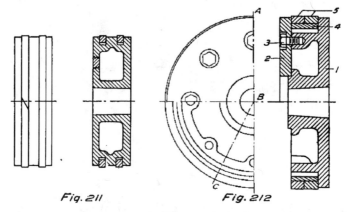

Fig. 211   Fig. 212

through the piston rod, crosshead, and connecting rod to the crank. The crank causes the shaft to revolve, carrying with it the flywheel, from which power may be transmitted by means of a belt.

Pistons. — Pistons are used in many forms of machines and vary accordingly. Some forms are shown in Figs. 211 and 212. The names of the parts for the form of steam piston shown in Fig. 212, are

1. Piston Body,
2. Follower,
3. Follower Bolts,
4. Bull Ring,
5. Packing Rings.

To prevent loss of pressure by leakage past the piston some form of packing ring is generally employed. Pistons are most always made of cast iron as are the rings. The rings are turned to a slightly larger diameter than the cylinder. A piece is then cut out and the ring is then sprung into place. For water pistons

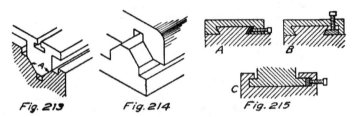

Fig. 213   *Fig. 214*   *Fig. 215*

a soft packing of hemp, fiber, or leather is used. For large vertical engines steel pistons are sometimes used.

Sliding Bearings. — Sliding bearings are of many forms, as shown in the following figures. The general end sought is to have the projected area of slide such that the pressure will not force out the lubricant and allow the metals to come into contact with each other. Smoothness of surfaces is only relative and

Fig. 216   *Fig. 217*   *Fig. 218*

surfaces in contact wear rapidly, hence the necessity for efficient lubrication.

Fig. 213 shows a form of planer guide. It is self-adjusting for wear and can be easily oiled. There is, however, considerable pressure between the inclined surfaces, which means that the power for operating the table increases as the angle A is decreased, and also the wear. A is commonly made 90° or less for small planers, while for heavy planers it may be 110° or more. The side pressure of the tool must be considered in selecting the proper value of A since it exerts a tendency to raise the table from the ways.

Fig. 214 shows the form generally used for lathe ways. It is self-adjusting, does not readily hold chips or dirt, but is not so easily kept oiled as Fig. 213.

There are many other forms of such bearing surfaces, some of which are provided with gibs for adjusting, as in Fig. 215. Com-

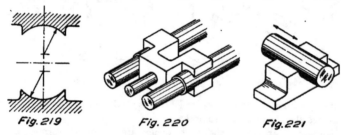

Fig. 219   Fig. 220   Fig. 221

mon forms of crosshead guides for steam engines are shown in Figs. 216, 217, and 218. Fig. 218 is used on all sizes of engines, and is satisfactory, since it allows the crosshead to adjust itself to the crank pin and connecting rod if turned concentric with the

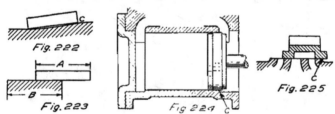

Fig. 222

Fig. 223   Fig. 224   Fig. 225

cylinder. Sometimes, however, the guides are turned with centers as in Fig. 219. This prevents turning.

For small pressures the form shown in Fig. 220 is often used, sometimes with one rod only. Fig. 221 is another form of sliding bearing. The pressure per square inch of projected area on crosshead guides should not exceed 100 pounds per square inch and may well be kept as low as 40 pounds per square inch.

Wear and Pressure. — Where there is much wear care must be used in the design of a sliding bearing and guide. Provision should always be made for running over at the ends of the guide. The same applies to the width of the guide. The effect of guides which are too long is shown much exaggerated by the shoulder

"*C*" in Fig. 222. Fig. 223 shows the correct design in which the slide runs over the guide at each end and causes more even wear. If "*A*" and "*B*" are made of equal length there will be equal wear. This same principle is involved in the piston and cylinder of a steam engine which accounts for the counterbore over which

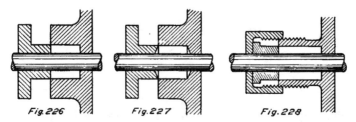

Fig. 226   Fig. 227   Fig. 228

the piston runs, "*C*" (Fig. 224), and similarly for slide valve seats (Fig. 225).

Stuffing Boxes. — Some common forms of gland and screw stuffing boxes used on engines, pumps, etc., for preventing leakage of steam or water around the piston rod where it passes through the end of the cylinder are shown in Figs. 226, 227, and 228. For rods $1\frac{1}{4}$ inch in diameter or less the common screw stuffing

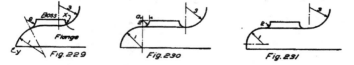

Fig. 229   Fig. 230   Fig. 231

box, Fig. 228, may be used. They are generally made of composition although they are sometimes made of cast iron for cheap work. The gland stuffing box (Figs. 226 and 227) is used for rods $1\frac{1}{2}$ inch and more in diameter. The box should be deep enough for four strands of packing and the gland so constructed as to be able to compress it to about one half its original size. These glands may have the bottom of the gland and box beveled as shown in Fig. 227. They may be lined with composition in which case the lining should be at least $\frac{3}{16}$ inch thick, but for rods less than $2\frac{1}{2}$ inch diameter it is generally advisable to make the gland entirely of composition. These are the common forms, but the student will do well to investigate some of the various types of metallic packings, since they are largely used in good designs.

Useful Curves and Their Application. — There are many small details in the actual drafting of a design which often give trouble out of proportion to their apparent importance when first en-

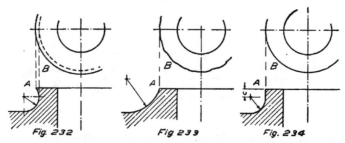

Fig. 232   Fig 233   Fig. 234

countered. The following suggestions are made to facilitate the drafting part of design, and not as rules to be strictly adhered to. Various curves which are commonly used are shown.

Fillets and Rounds. — The drawing of fillets and quarter rounds deserves attention, since they are of so frequent occurrence. Fig. 229 shows a portion of a machine. The centers and radii of the various arcs are indicated. All radii are too large, but

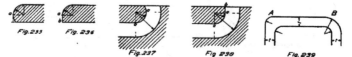

Fig. 235   Fig. 236   Fig. 237   Fig 238   Fig. 239

especially 1 and 2. Radius 1 gives a point at y. Radius 2 is so large that it cannot be used for the complete circumference of the boss as indicated at x. Of course a changing radius of fillet might be used, but this would not allow the use of ready made fillet strips. Fig. 230, in which the limiting radii are used, is an improvement. Fig. 231 shows a much better design. Note that the radii 1 and 2 are less than the thickness of the flange and boss respectively. The effect of a quarter circle is obtained by this method in which the flange and boss each start with a straight line. The straight line also produces a better appearance after finishing off the surface of the boss. This is shown in Figs. 232, 233, and 234, where the effect of different fillets is indicated at B

in each of the views. In the first case there is an undercutting, in the second view *B* shows the irregular outline produced, while the third case shows the correct design.

Arcs and Straight Lines. — When arcs are used in connection with straight lines the fault shown at *a* in Figs. 235 and 237 should be avoided. Do not run the arc past the tangent point "*a*", and notice that the line *a-b* is a straight line in Figs. 236 and 238.

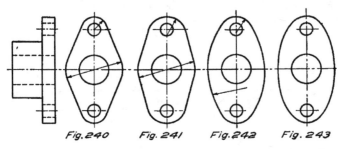

Fig. 240   Fig. 241   Fig. 242   Fig. 243

At *A* in Fig. 239 is shown the effect of not changing the radius when two parallel lines are continued by arcs. At *B* the thickness of material has been kept by maintaining the same center and changing the radius by the distance *t*.

Flanged Projections. — When flanged projections are used with bolts or nuts they may take a variety of shapes, some of

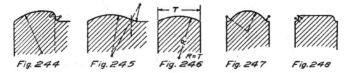

Fig. 244   Fig. 245   Fig. 246   Fig. 247   Fig. 248

which are shown in Figs. 240 to 243. After locating the centers of the bolt holes the extent of the flange may be found by adding twice the bolt diameter to the distance between bolt centers. Frequently the outline is obtained as in Fig. 240 in which an arc is drawn from the center of the bolt hole with a radius equal to the diameter of the bolt.

A much better appearance is obtained by using a larger radius whose center is at the intersection of the bolt hole and the center line, as shown in Fig. 241. Either straight or curved lines may be used to join the small and large arcs. Sometimes an ellipse may be used. A gland is used for illustration, but similar cases occur in pipe connections, the bolted feet of machines, etc.

Flange Edges. — Flanges are often finished with curves so as to avoid machining. Several forms are shown in Figs. 244 to 248. The radius R may be taken equal to the thickness T. The centers for the various radii are indicated.

Flanges and Bolting. — A method of finding the diameter of bolt circle and diameter of flange is illustrated in Figs. 249, 250, and 251. For through bolts consider Figs. 249 and 250. Draw

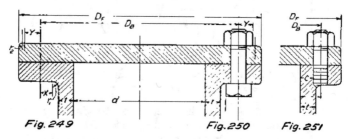

Fig. 249 Fig. 250 Fig. 251

in a proper fillet at r_1. For a trial the radius r_1 may be taken as one fourth of the thickness of the cylinder wall t. Then lay off X, equal to one half the distance across flats of bolt head, and Y, equal to one half the distance across corners of nut. The diameter of the bolt circle, D_B, may now be found by laying a scale on the drawing and selecting a dimension. This will be equal to, or greater than, $d + 2(t + r_1 + X)$, and may be taken at the nearest $1/8$th inch. The flange diameter may then be obtained by laying out the distance Y, as in Fig. 249, and using the scale to find an even dimension equal to, or greater than, $D_B + 2(Y + r_2)$. The radius r_2 may be taken at $1/8$th to $1/16$th the thickness of the flange. When studs are used the diameters D_B and D_F may be greatly decreased as shown in Fig. 251. The distance C should be about equal to t, although if necessary it can be made equal to one half the diameter of the bolt.

Keys. — Keys of various forms are used to prevent relative motion between shafts and pulleys, gears, crank arms, etc. The common forms are here shown. Fig. 252 is called a saddle key and may be used where only a small force is to be transmitted and where close or frequent adjustment is required. Fig. 253 is called a flat key, and requires a flat spot upon the shaft. Its holding power is a little greater than the preceding form. Set screws are sometimes used with Figs. 252 and 253 to secure a closer contact. Fig. 254 is the most common form, and may be either square or rectangular in section. The sides of the key should fit closely in the hub and shaft. Various proportions are given for keys. Square keys are often made with

$$W = \frac{D}{4}$$

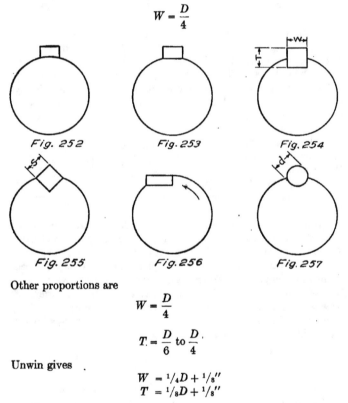

Fig. 252   Fig. 253   Fig. 254

Fig. 255   Fig. 256   Fig. 257

Other proportions are

$$W = \frac{D}{4}$$

$$T = \frac{D}{6} \text{ to } \frac{D}{4}$$

Unwin gives

$$W = \frac{1}{4}D + \frac{1}{8}''$$
$$T = \frac{1}{8}D + \frac{1}{8}''$$

The taper for keys may be from $\frac{1}{16}$th to $\frac{3}{16}$ths of an inch per foot of length. One eighth inch is often used. The key should be half in the shaft and half in the hub. When the force to be transmitted is very large two keys may be used. In such cases they are generally placed 90° apart. The length of keys

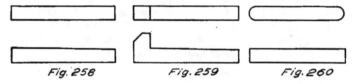

Fig. 258   Fig. 259   Fig. 260

should be one and one half or more times the diameter of the shaft. Fig. 256 shows the Lewis key, invented by Wilfred Lewis. The direction of rotation for the driving shaft is indicated. It will be noted that this form is wholly under compression. Fig. 255 is a different way of locating a square key. The side S may be taken as one fourth the diameter of the shaft. Fig. 257 shows a round key. It is a desirable form when it can be used, as when located at the end of a shaft. Fig. 258 shows the ordinary plain key; Fig. 259, a key provided with a gib to make its removal easier. Fig. 260 shows a round end

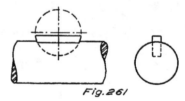

Fig. 261

key which may be fitted into a shaft. Such keys are often used when it is desired to arrange for a part to slide on the shaft. When a long key is secured in a shaft and used for this purpose it is called a feather or feather key. Square end keys may be used in the same way. Fig. 261 shows the Woodruff Key, which consists of a part of a circular disc. They are made in a variety of sizes with dimensions suiting them to different purposes. The circular seating allows the key to assume the proper taper when a piece is put onto the shaft.

CHAPTER XIII

SKETCHING

Uses of Sketching. — Freehand sketching is of particular importance in connection with drafting and will be briefly considered in this chapter. All that has been said in the previous chapters concerning the theory and practice of drafting applies to freehand sketching. The term sketching must not be considered as indicating incompleteness, for if anything a sketch must be more complete than a mechanically executed drawing. Sketching is the engineering language of the trained executive as well as a convenient and quick method of representation. Sketches are used to give information from which parts are to be made; they are used for repair parts; new parts; as an aid to reading drawings; as an aid to design; as a means of recording ideas, and for many other purposes.

Accuracy of thought, observation, representation, and proportion are essential. The four *"P's"* of sketching are *practice, patience, proportion, and proficiency.* Too much emphasis cannot be put upon the necessity of accuracy in proportion and detail.

A most interesting example is shown in Fig. 262 which is a reproduction of a sketch for the first steam hammer as drawn by James Nasmith. Quoting from Nasmith's autobiography by Samuel Smiles: * "I got out my 'scheme book,' on the pages of which I generally *thought out*, with the aid of pen and pencil, such mechanical adaptations as I had conceived in my mind, and was thereby enabled to render them visible. I then rapidly sketched out my steam hammer, having it all clearly before me in my mind's eye. In a little more than half an hour after receiving Mr. Humphrie's letter, narrating his unlooked-for difficulty, I had the whole contrivance in all its executant details, before me in a page of my scheme book. The date of this first drawing was November 24, 1839."

* Published by Harper and Bros., New York.

Materials for Sketching. — The materials necessary for sketching are a 2H drawing pencil, pencil eraser, art gum, and paper. Either plain or squared paper may be used, but it is better to use the plain paper at first so as not to be dependent upon the aid

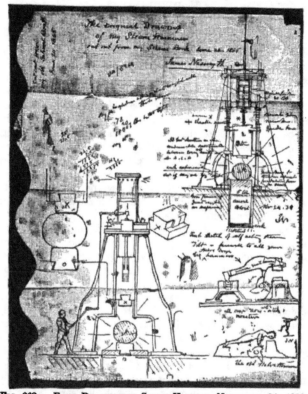

FIG. 262. — FIRST DRAWING OF STEAM HAMMER, NOVEMBER 24, 1839.

which the squares give. The pencil should be kept well sharpened with a long round point. It is desirable to have a small board on which the paper may be tacked, or clip boards such as are used by bookkeepers will be found very convenient as a means of holding the paper. Every sketch should have a title, the date, and the name of the person who made it.

Making a Sketch. — To make a sketch the following order may be pursued. First examine the object, determine the number of views necessary completely to define it, and observe the proportions. Then proceed to sketch very lightly, locating center lines and blocking in the limits for all views. Sketch in the details and then go over and brighten up wherever necessary in order to make all parts clear and definite. Straight lines may be drawn by making a succession of short straight lines or by

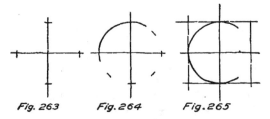

Fig. 263   Fig. 264   Fig. 265

marking points and drawing from one point to another. Views should be blocked in completely with straight lines regardless of the number of curves and circle arcs.

To sketch a circle draw center lines at right angles (Fig. 263), space off radii, as shown in Fig. 264, on the center lines and in between them. Another method is to block in a square made up of four smaller squares (Fig. 265), then sketch in one fourth of the required circle at a time.

Taking Measurements. — There are a great many tools used for determining the sizes of machine parts and constructions. The names of some of the tools should be learned together with the methods of using them and the conditions under which they are used. For this purpose the reader is advised to secure a catalog of machinist's tools. Some of the tools used for various purposes are:

The two foot rule for comparatively rough work.

The standard steel rule for more accurate work. It should have both binary and decimal divisions.

Steel tapes used for measuring rather long distances.

Straight edge, used for extending surfaces.

The square, used in a variety of forms; fixed, adjustable, combination.

Calipers, used for obtaining distances.   There are many forms; outside, inside, spring, transfer.
  Surface plate and surface gage.
  Depth gage and hook gage or scale.
  Plumb bob.
  Micrometer.
  Vernier caliper.
  Plug and ring gages.
  Wire and sheet metal gages.
  Screw thread gages.
  Radius gages.

The surfaces to be measured are flat surfaces and curved surfaces. These will appear in many combinations and will require separate consideration in each case. Cylinders may be measured directly with the calipers or scale. A steel tape may be used to measure the circumference of a large cyl-
inder and the diameter
calculated.   Angular
measurements are
made with some form

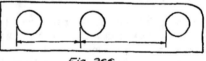

Fig. 266

of protractor. The bevel protractor and center square are useful for this purpose. The use of chalk or a marking solution is often necessary or convenient. Curved outlines may be obtained by offset measurements, by rubbing an outline on paper, or by making a template by such means as the conditions permit. Center distances may be found by measuring from the edge of one hole to the corresponding edge of the next hole as indicated in Fig. 266.

The question of accuracy in taking measurements will arise frequently. The finished or machined parts should be measured as accurately as the means at hand will allow. Shafts or sliding blocks, or wherever a fit is involved, should be measured with the micrometer or similar accurate means. Rough castings of small or medium size may be measured to the nearest $1/16$th inch, while larger ones may be near enough when measured to $1/8$ or even $1/4$th inch. In all cases judgment must be exercised, and whenever in doubt take measurements as closely as possible under the conditions.

Where the parts being sketched are for repairs or replacement, very accurate measurements are often required, and in the case of a fit allowance for wear must be made.   If a whole new machine or construction is to be built much time can often be saved by less accurate measurements, as the parts will be dimensioned to go together when the final drawing is made.   Ingenuity and common sense are the primary requisites.

In connection with measurements it will be necessary to know something of standard nomenclature.   For instance, the three

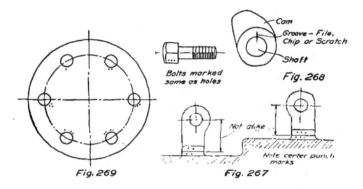

Bolts marked
same as holes

Cam
Groove – File,
Chip or Scratch
Shaft
Fig. 268

Not alike

Note center punch
marks

Fig. 269   Fig. 267

dimensions of a taper are indicated by a single number and a name.

Some Ideas on Sketching. — The difficulties which are to be met and overcome when making sketches under trying circumstances with limited time, inaccessability, with a machine in operation in close quarters, etc. — is little understood or appreciated by those accustomed to the conveniences of the drafting room.

Many times sketches are made only for one's own use and so can perhaps be made a little less presentable than when made to take the place of a drawing.   However, there is a warning which must be sounded, and that is the unvarying rule "to preserve definiteness under all circumstances."   A sketch may be hastily made, but a careless sketch is worse than useless.   Be sure that what is given is right and of *certain* meaning.   The steps which must be followed in making a sketch are:

Sketch the parts.

Put on dimension lines and notes.

Measure the parts and fill in the figures.

Some considerations to be kept in mind are: —

Use part views to show special features or details.

Use notes freely but not as a substitute for necessary views.

Show hexagons, octagons, etc., across flats using a note to tell the number of sides or insert a revolved section.

Note identification marks, and mark parts to facilitate putting them together and for fixing relative positions.

Note finished surfaces and kinds of finish.

Use templates whenever in doubt as to curves, location of drilling, etc.

Note materials of which machine or parts are made.

Measure sizes of holes as well as of bolts, shafts, etc.

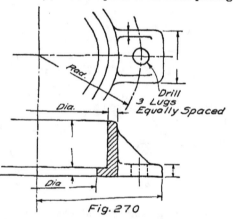

Fig. 270

A small amount of surface shading is often of value.

Note the location of the machine in reference to other machines or to building features if such information has any possibility of being useful.

Rods, bolts, bars, and long pieces of uniform section can generally be shown in one view.

Most machines and some parts of machines will carry the manufacturer's name and identification, sometimes stamped into the machine, and sometimes on a name plate. The information given in this manner should always be noted in connection with the sketch. Sometimes parts are either right or left hand, and this fact should be noted. It is a good plan to examine all parts very carefully for identification marks.

When parts bear a definite relation to one another, prick punch marks or a filed groove will often be of great assistance in reassembling (Figs. 267 and 268). Oftentimes the top or bottom of a part should be marked. Where a number of bolts are used with reamed holes they are often numbered or otherwise marked (both bolt holes and bolts, Fig. 269). Very often part views may be used to save time by adding a note: For instance, a circular object with lugs, as shown in Fig. 270. In the case of cylindrical objects the word "diameter" will often save a view. A washer

Fig. 271

would be sketched as in Fig. 271. Sections are rather freely used in sketching as they give prominence to the sketch. It is often desirable to make a separate outline sketch without dotted lines in connection with a sectional drawing of a part, especially when the sketches must be hastily made, as the two sketches result in less confusion than when combined in one view.

When sketches are made in connection with diagrams for the transmission of power, or a mechanism of any sort, the computations should be included with the sketch, and existing pulleys or other parts should be clearly dimensioned and indicated to distinguish them from proposed additions. In the case of foundations where bolts are to be located, differences in level

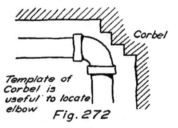

Template of Corbel is useful to locate elbow   Fig. 272

must be considered as well as center line distances. When locating shaft hangers, or constructions to be fastened to a wall or ceiling, the surroundings such as parts of the permanent structure, like beams or corbeling of the brick wall (Fig. 272), should be measured and sketched with the part to be installed.

The principal point to be brought out in connection with sketching of any kind is to leave nothing to guess — to have too much rather than too little information, and to make every line and note absolutely definite.

CHAPTER XIV

ESTIMATION OF WEIGHTS

Accuracy. — It is often necessary to compute the weight of machine parts or of piles of materials; for instance, to estimate the amount of coal on hand. The annual stock taking of many companies requires much of this work which must be accomplished accurately and expeditiously. Some of the methods used should be known together with the degree of accuracy required. For some purposes a result within 5 % or even 10 % may be sufficiently close, while in other cases an accurate result may be desirable, as when figuring a large number of pieces of expensive material. The weights of many standard parts are well known and are given in manufacturers' catalogs. The weights of steel shapes are known and tabulated in pounds per linear foot, the weight of bolts per 100, and similarly for other pieces.

Weights of Materials. — The following weights are average values for various materials and may be used for ordinary calculations.

Material	Pounds per Cubic Inch	Pounds per Cubic Foot
Cast Iron	.26	450
Wrought iron	.28	480
Steel	.29	490
Brass	.30	530
Copper	.32	550
Lead	.41	710
Aluminum	...	160
Granite	...	170
Brick	...	120
Concrete	...	145
Water	.036	62.5
Spruce	...	30
White pine	...	30
Yellow pine	...	41
Maple	...	45
Lignum vitae	...	83
Oak	...	50

Weight of Loose Materials. — In estimating the amount of material in a pile, its shape may be approximated to one or more geometrical forms and its volume computed. This is best done by making a sketch with dimension lines which are filled in with measurements. Such sketches should be preserved for checking purposes and as a record. The weight per cubic foot or yard is then obtained by loading a car of measured volume and weighing it or by filling a box containing a cubic foot or yard and finding the net weight. The material should of course be disposed as

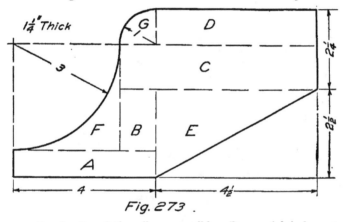

Fig. 273 .

near the density of the pile as possible. By careful judgment and some experience a very close approximation of weight may be obtained in this manner. For more accurate work, the surveyor's transit may be used.

Weight of Castings. — The computation of the weight of castings most frequently occurs either in connection with the cost or where a machine must come within certain limits of weight. The weight may be calculated from the drawings. For simple objects this is not difficult, but for many shapes much loss of time may be saved by systematic methods and proper division into elementary forms. Two sets of weights must be considered; one the object in the rough, and the other the finished piece. Allowances for finish must be made. It is necessary to know what holes or openings are to be cored and what ones are to be machined. Cylindrical pieces are readily figured by dividing into separate cylinders. Limits as to weight are very important when machines must be assembled in out of the way places, or where transportation is by pack mules or other primitive means.

Methods of Calculation. — The general method of finding the weight of a piece is to compute its total volume in cubic inches and then multiply this volume by the weight of a cubic inch of the material. Most pieces may be divided into flat plates, cylinders, and flanges, each of which should be lettered and tabulated. Sometimes fillets may be balanced against bolt holes or against rounded corners. In other cases the fillets may be considered as a certain per cent of the whole. The weight as figured should also be increased to allow for rapping the pattern in the mold. The allowance for finish may be $1/8''$ for general work but this varies with different classes of work and with the degree of accuracy required in the finished piece.

When a piece has a uniform thickness but irregular outline it may be broken up into plane figures and the area of each found separately (Fig. 273). After adding them together multiply by the thickness to obtain the volume and then by the unit weight to find the total weight, as illustrated. The dash lines divide the flat surface into seven parts, each of which is lettered. These may be listed in tabular form.

Designation	Part	Dimensions Inches	Area Square Inches
A	Rectangle	$4 \times 3/4$	3.
B	"	$1^3/4 \times 1$	1.75
C	"	$5^1/2 \times 1^1/4$	6.875
D	"	$4^1/2 \times 1$	4.5
E	Triangle	$1/2(2^1/2 \times 4^1/2)$	5.625
F		$1/4(36 - 28.27)$	1.93
G	Circle	$1/4(3.1416)$	.785

Total area square inches 24.465

Volume = area × thickness

= 24.47 × 1.25 = 30.59 cubic inches

The area of part G is one fourth the area of a circle having the radius indicated. The area of part F is found by subtracting one fourth the area of a circle having the radius given from the area of a square, one side of which is equal to the radius of the arc.

With irregular shapes the area is sometimes divided approximately into regular figures, the dimensions for which are obtained by applying the scale to

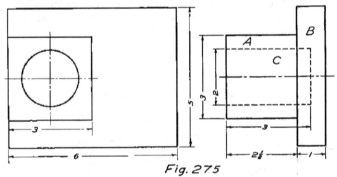

the drawing. This is illustrated in Fig. 274 where the dash line x-x is drawn so that the area B appears to be equal to the area A + A. The distance H is then measured and multiplied by L to find the area. In the case of hollow pieces, find the volume as though the piece was solid, then subtract the volume of the spaces.

Fig. 275

In Fig. 275 the volume would be found as tabulated, in which the A and B are called plus (+) volumes and C is called a minus (−) volume.

Designation	Part	Dimension	Volume in Cubic Inches	
			+	−
A	Square prism	$3 \times 3 \times 2^1/_2$	22.5	
B	Rectangular plate	$5 \times 6 \times 1$	30.	
C	Cylinder	$\cdot\ 1 \times 3.1416 \times 3$		9.42
Totals...			52.5	9.42

$(A + B) − C =$ Net volume
$52.5 − 9.42 = 43 \pm$ cubic inches

For the ring shown in Fig. 276 find the area of the cross section A
and multiply by the circumference of the mean diameter. This
method is often a convenient one.

Weight of Cylinder Head. — To find the approximate weight
of the small cylinder head of Fig. 277 it may be divided into

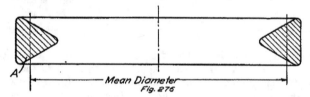

Fig. 276

three cylinders, two positive and one negative. The round at
x may be balanced against the fillet at y for approximation purposes. Allow say $1/16$th inch on each of the finished surfaces.
The calculations will be as tabulated.

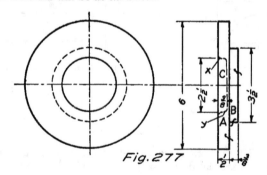

Fig. 277

Designation	Part	Dimensions Inches	Volume Cubic Inches	
			+	−
A	Cylinder	$28.27 \times {}^9/_{16}$	15.90	
B	"	$9.62 \times {}^3/_8$	3.61	
C	"	$4.91 \times {}^3/_8$		1.84
Total...			19.51	1.84

$(A + B) - C$ = net volume
$19.51 - 1.84 = 17.67$ cu. in.
Vol. × wt. per cu. in. = total weight
$17.67 \times .26 = 4.60$ pounds

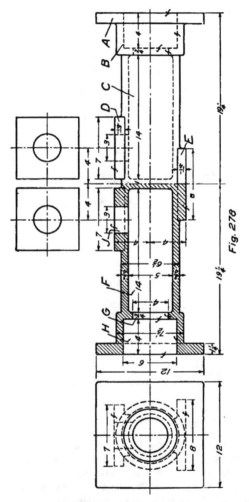

Weight of Plunger Barrel. — To approximate the weight of the pump barrel shown in Fig. 278. First divide it into parts as indicated in the figure. The plus volume treats it as a solid. The minus volume consists of the interior cylindrical spaces

H, G, F, and J. The calculations for its cost at ten cents per pound follow. For any other price multiply by the required cents per pound and divide by ten. Since both ends are alike only one half is figured and the result is then multiplied by two.

Designation	Part	Dimensions Inches	Volume Cubic Inches	
			+	−
A	Flange	$12 \times 12 \times 1^1/_4$	180	
B	Stuff box	$\dfrac{\pi(7.5)^2}{4} \times 3^1/_2$	154	
C	Main cylinder	$\dfrac{\pi(6.5)^2}{4} \times 14^1/_2$	481	
D	Port flange	$7 \times 7 \times 1^1/_4$	62	
E	Foot flange	$4 \times 8 \times 1^1/_4$	80	
F	Cylinder	$\dfrac{\pi(5)^2}{4} \times 14$		274
G	Throat	$\dfrac{\pi(4)^2}{4} \times {}^3/_4$		9
H	Stuff box	$\dfrac{\pi(6)^2}{4} \times 4$		116
J	Port	$\dfrac{\pi(3)^2}{4} \times 1^1/_2$		11
Total volumes.............................			957	410
Multiplied by 2 for two ends................			1914	820

1094 cu. in. net volume
1094 × .26 = 285, pounds weight
285 × .10 = $28.50, cost of casting at 10 cents per pound

Weight of Forgings. — Steel and wrought iron shafts may be readily figured, especially when turned from stock bars or rods. Forgings, however, require careful consideration as the rough forging may weigh from 25 % to 50 % more than the finished piece, especially if the shape is at all complicated.

CHAPTER XV

PIPING

Piping Materials. — Pipe made of various materials is used for conveying liquids and gases. For a complete treatment of the subject of piping and its uses, piping drawings, etc., see the author's "Handbook on Piping," D. Van Nostrand Company, N. Y. The illustrations for this chapter are from the above book.

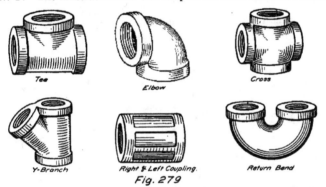

Tee   Elbow   Cross

Y-Branch   Right & Left Coupling.   Return Bend

Fig. 279

Cast iron pipe is cheaply made and is used for underground gas, water, and drain pipes, sometimes for steam and exhaust pipes where low pressures are carried.

Wrought iron or steel pipe is most commonly used, especially where high pressures are encountered. Copper is used to a certain extent where there is limited room. For hot water or bad water, brass pipe is to be preferred as it does not corrode like iron or steel. Spiral riveted steel piping is often used for large pipes.

Pipe Fittings. — For joining lengths of pipe and making turns and connections, "fittings" are used, Fig. 279. Such fittings consist of flanges, couplings, tees, ells, crosses, etc. Small pipe is often "made up" by means of couplings and screwed fittings — large sizes use flanges and flanged fittings. Some general information is given in the tables included in this chapter.

112

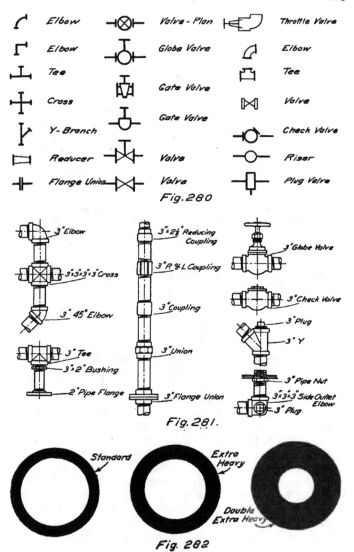

⌐	Elbow	Valve - Plan			Throttle Valve
Γ	Elbow	Globe Valve			Elbow
⊥	Tee				Tee
+	Cross	Gate Valve			Valve
Y	Y-Branch	Gate Valve			Check Valve
⊐	Reducer	Valve			Riser
+	Flange Union	Valve			Plug Valve

Fig. 280

3" Elbow
3"x3"x3"x3" Cross
3" 45° Elbow
3" Tee
3"x2" Bushing
2" Pipe Flange

3"x 2½" Reducing Coupling
3" R. & L. Coupling
3" Coupling
3" Union
3" Flange Union

3" Globe Valve
3" Check Valve
3" Plug
3" Y
3" Pipe Nut
3"x3"x3" Side Outlet Elbow
3" Plug.

Fig. 281.

Standard   Extra Heavy   Double Extra Heavy

Fig. 282

The representations of Figs. 280 and 281 are often used when making piping layouts.

Standard Pipe. — Wrought pipe is known by its nominal inside diameter. In the United States the Briggs Standard is in general use. The nominal diameter differs from the actual diameter by varying amounts, as indicated in the Table. Standard pipe is

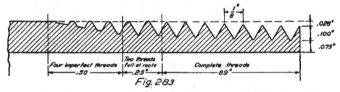

Fig. 283

used for pressures up to 125 pounds per square inch. Extra strong and double extra strong pipe are made for use at higher pressures. The extra thickness is obtained by reducing the inside diameter, the outside diameter remaining constant for a given nominal diameter. The actual cross sections for the three weights of $^3/_4$ inch pipe are shown in Fig. 282.

Pipe Threads. — Pipe threads are cut with an angle of 60°, with the top and bottom rounded, making the height .8 of the pitch. The threads are also cut on a taper of three fourths inch per foot as illustrated in Fig. 283.

DIMENSIONS OF STANDARD WROUGHT PIPE

Nominal Diameter, Inches	Actual Inside Diameter, Inches	Actual Outside Diameter, Inches	Threads per Inch	Length of Perfect Thread, Inches
$^1/_8$	.269	.405	27	.19
$^1/_4$	.364	.540	18	.29
$^3/_8$	.493	.675	18	.30
$^1/_2$	.622	.840	14	.39
$^3/_4$	.824	1.050	14	.40
1	1.049	1.315	$11^1/_2$	.51
$1^1/_4$	1.380	1.660	$11^1/_2$	.54
$1^1/_2$	1.610	1.900	$11^1/_2$	.55
2	2.067	2.375	$11^1/_2$	.58
$2^1/_2$	2.469	2.875	8	.89
3	3.068	3.500	8	.95
$3^1/_2$	3.548	4.000	8	1.00
4	4.026	4.500	8	1.05

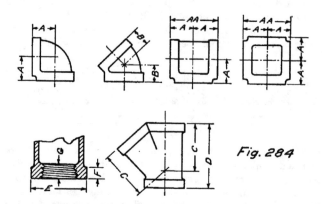

Fig. 284

DIMENSIONS OF WALWORTH MFG. CO. CAST IRON FITTINGS

Size of Pipe, Inches	A Inches	A-A Inches	B Inches	C Inches	D Inches	E Inches	F Inches	G Inches
$1/4$	$3/4$	$1 1/2$	$7/16$	...	...	1	$1/4$	·$3/8$
$3/8$	$7/8$	$1 3/4$	$9/16$	$1 7/16$	$2 1/16$	$1 1/8$	$5/16$	$7/16$
$1/2$	$1 1/16$	$2 1/8$	$11/16$	$1 7/8$	$2 9/16$	$1 7/16$	$3/8$	·$1/2$
$3/4$	$1 5/16$	$2 5/8$	$13/16$	$2 1/16$	$2 3/4$	$1 3/4$	$7/16$	$9/16$
1	$1 1/2$	3	$15/16$	$2 1/2$	$3 1/4$	$2 1/16$	$1/2$	$5/8$
$1 1/4$	$1 13/16$	$3 5/8$	$1 1/16$	3	$3 3/4$	$2 1/2$	$9/16$	$11/16$
$1 1/2$	2	4	$1 3/16$	$3 1/4$	$4 3/4$	$2 3/4$	$5/8$	$13/16$
2	$2 3/8$	$4 3/4$	$1 3/8$	4	$5 1/2$	$3 3/8$	$11/16$	$7/8$
$2 1/2$	$2 7/8$	$5 3/4$	$1 5/8$	5	$6 13/16$	$4 1/8$	$13/16$	1
3	$3 5/16$	$6 5/8$	$1 7/8$	$5 5/8$	$7 5/8$	$4 3/4$	$15/16$	1
$3 1/2$	$3 11/16$	$7 3/8$	$2 1/16$	$6 3/8$	$8 3/4$	$5 1/4$	1	$1 1/16$
4	4	8	$2 1/4$	$7 1/8$	$9 3/4$	6	$1 1/16$	$1 1/8$

AMERICAN STANDARD PIPE FLANGES

125 Pounds Working Pressure

Pipe Size, Inches	Diameter of Flange, Inches	Thickness of Flange, Inches	Diameter of Bolt Circle, Inches	Number of Bolts	Diameter of Bolts, Inches
1	4	$7/16$	3	4	$7/16$
$1^1/_4$	$4^1/_2$	$1/_2$	$3^3/_8$	4	$7/16$
$1^1/_2$	5	$9/16$	$3^7/_8$	4	$1/_2$
2	6	$5/_8$	$4^3/_4$	4	$5/_8$
$2^1/_2$	7	$11/16$	$5^1/_2$	4	$\cdot5/_8$
3	$7^1/_2$	$3/_4$	6	4	$5/_8$
$3^1/_2$	$8^1/_2$	$13/16$	7	4	$5/_8$
4	9	$15/16$	$7^1/_2$	8	$5/_8$
$4^1/_2$	$9^1/_4$	$15/16$	$7^3/_4$	8	$3/_4$
5	10	$15/16$	$8^1/_2$	8	$3/_4$
6	11	1	$9^1/_2$	8	$3/_4$
7	$12^1/_2$	$1^1/16$	$10^3/_4$	8	$3/_4$
8	$13^1/_2$	$1^1/_8$	$11^3/_4$	8	$3/_4$

CHAPTER XVI

INTERSECTIONS

The Line of Intersection. — The line of intersection of two surfaces is that line which contains all the points which are on both of the surfaces. Objects in general are made up of parts and where these parts come together there is said to be a line of intersection, as shown in Figs. 285 and 286. The chimney intersects the roof and there is also an intersection between the dormer window and the roof. The intersection between two cylinders is shown in Fig. 286.

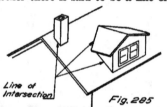

Line of Intersection

Fig. 285

It is often necessary to determine the intersection of two surfaces, either to find the appearance or for purposes of development.

The intersection between two planes is a straight line as shown in Fig. 287. If these planes cut a cylinder or cone the lines of

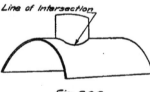

Line of Intersection

Fig. 286.

intersection may be straight or curved (Figs. 288 and 289). If the plane is at right angles to the axis a right section is cut as shown by the horizontal planes which intersect the cylinder and cone in circles. If the plane passes through the axis it intersects the cylinder in a straight line parallel to the axis called an element. In like manner an element may be cut from the cone. Note that all the elements of a cylinder are parallel, and that all the elements of a cone pass through the apex.

Intersecting planes, elements, and cut sections are the basis for finding lines of intersection of surfaces.

117

Intersection of a Vertical Prism and a Horizontal Prism. — Fig. 290 shows a square prism intersecting a triangular prism. Two methods of solution may be used. First method: Examine the three views, then note that the top view shows where the

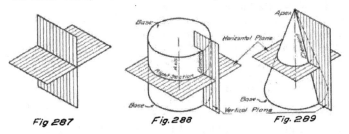

Fig 287   Fig. 288   Fig. 289

edge AB of the square prism pierces the front face of the triangular prism at point B^{π}. The front and side views of this point may be obtained by projection and are shown at B^{v} and B^{s}. Note that the front view shows the intersection of the edge EF

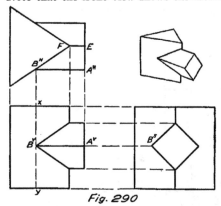

Fig. 290

of the square prism with a vertical edge of the triangular prism. Project to the other views. Join the points thus found which will determine the projections of a line of intersection between the two prisms. Second method: Imagine a vertical plane to be passed through the edge AB. This plane will intersect the face of the triangular prism in a vertical line xy shown in the front view. Since the lines xy and AB are in the same plane, the point in which they cross will show in the front view at B^{v}. By passing similar planes through each of the edges the other points may be found.

Intersection of a Vertical Prism and an Inclined Prism — Visibility of Points. — The intersection of two prisms, one of which is inclined, is shown in Fig. 291. Either of the methods just described may be used, but the second method is to be pre-

ferred. A cutting plane must be passed through each edge of both prisms within the limits of the curve of intersection. This means all of the edges of either prism through which a plane may be passed that will cut the other prism. A plane passed through the front edge of the vertical prism would not cut the inclined prism, and so would not locate any points on the line of intersection. A vertical plane through line AB will intersect the front face of the rectangular prism in line C^vD^v. The point in which these lines cross is shown in the

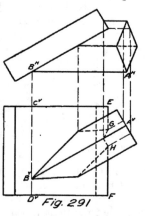

Fig. 291

front view at B^v. Since *both* lines are on *visible* faces of the prisms the *two* lines are visible and the point B^v is visible. Lines

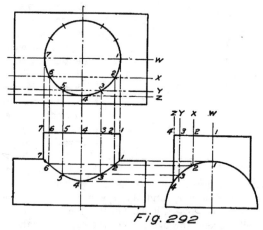

Fig. 292

of intersection in order to be visible must join two visible points determined as stated. A vertical plane through the edge EF will intersect the inclined prism in two lines parallel to the inclined

edges as shown. Each of these inclined lines intersects the edge
EF so that the two points *G* and *H* are located. The edge *EF*
would be visible if the inclined prism was not in front of it. The
two inclined lines, however, are on the back or invisible faces of
the inclined prism and so are invisible. The points *G* and *H* are

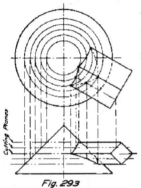

therefore invisible. A line joining
two invisible points or one visible
and one invisible point is invisible.
Lines which are visible in one view
may or may not be invisible in
another, and should be considered
separately.

Intersecting Cylinders. — Two intersecting cylinders are shown in Fig.
292. Divide the small cylinder into
equal parts and then pass planes
which will cut elements from both
cylinders. The planes *w*, *x*, *y*, and *z*
cut elements *1*, *2*, *3*, and *4* from the
cylinders. The points in which ele-

Fig. 293

ments in the same plane cross are shown in the front view at
points *1*, *2*, *3*, *4*, etc., thus determining the curve of intersection.
Use as many planes as are necessary to obtain a smooth curve.

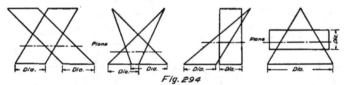

Fig. 294

Be sure to pass planes through the contour or outside elements
of both cylinders in order to obtain the extreme limits of the
curve. This is very important, especially when the axes of the
cylinders do not intersect.

Choice of Cutting Planes. — Whenever possible planes should
be passed so as to cut straight lines from both surfaces. The
lines (not parallel) on the same plane intersect in points which
are common to both surfaces and are therefore points in the
curve of intersection. The intersection between surfaces can
very often be found by horizontal cutting planes, as indicated

in Fig. 293, which would be employed for the cases presented in Fig. 294 and similar conditions. Considering Fig. 293 it will be observed that horizontal cutting planes are used. Each plane cuts a straight line from the prism and a circle from the cone,

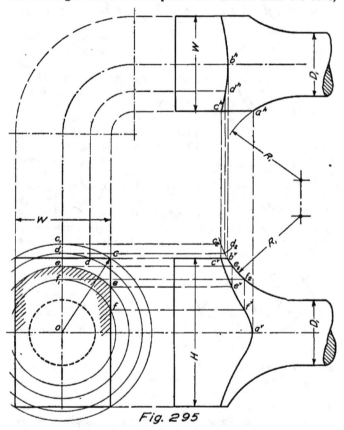

Fig. 295

as shown in the top view. Where the line and the circle cross is a point common to the prism and the cone. Other points found in the same way will complete the curve of intersection.

Connecting Rod Intersection. — Fig. 295 shows a portion of a connecting rod of circular cross section with a rectangular end.

The circular section is increased where it joins the rectangular portion. The curves of intersection are found as described. Notice that the centers for the radii R_1R_1 are in the same perpendicular line. D_1 is the diameter of the rod. There are certain "critical points" and these will be mentioned first. Where R_1 cuts the width of the rectangular part in the top view gives point a^h and this point will fall on the center line in the side view and so is projected to a^v. In a similar manner point b^v may be projected to the top view at point b^h. The end view is needed to obtain the other points. With O as a center and the corner distance OC as a radius, draw the arc CC_1. Continue the radius R_1 in the side view. A horizontal line through C_1 will intersect radius R_1 at C_2 from which C^v and C^h may be projected. The radius OC gives the largest circle which will touch the rectangular section and so determines one end of the curve, as shown. A plane passed through C^v or to left of point C^v and perpendicular to the axis will give a rectangular section. A plane to the right of point C^v will give other sections which will be described.

To determine the curve in top view. Two points b and c are already determined. For any other point d in end view, draw an arc dd_1, with od as a radius. From d_1 project horizontally to d_2 and then as shown to d^h in the top view.

To determine the curve in the side view. Two points a and c are already determined. Take any point e in the end view and with a radius oe draw arc ee_1; project horizontally from e_1 to e_2. The intersection of a vertical line through e_2 with a horizontal line through e will give point e^v, a point on the desired curve. Point f and other points are found in the same manner. It will be observed that a plane through e and perpendicular to the axis would give the section indicated by section lines in the end view.

CHAPTER XVII

DEVELOPMENTS

Surfaces. — Surfaces may be divided into two classes, plane surfaces and curved surfaces. Plane surfaces show in their true size and shape when they are parallel to one of the planes of projection, so that an object bounded by plane surfaces can have each of its faces brought into contact with a piece of paper, either by wrapping the paper about the object or placing the

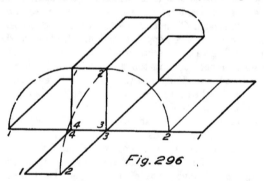

Fig. 296

object on the paper and then turning it until each face has touched the paper. This is shown in Fig. 296 where the paper has been cut so it will exactly cover the object when it is folded about it. Such an outline is called a development. A curved surface does not show in its true size no matter how it is placed with regard to the planes of projection. Some kinds of curved surfaces can be developed by rolling them on a plane as illustrated in Fig. 297. The distance L is equal to the distance around the cylinder and the height H of course remains equal to the length of the cylinder. Other surfaces, such as the surface of a sphere, cannot be exactly developed, but there are approximate methods which are generally accurate enough.

Development of a Prism. — The prism of Fig. 296 is developed by laying out in a straight line and in the proper order the distances *1–4*, *4–3*, *3–2*, and *2–1*, which added together are equal to the

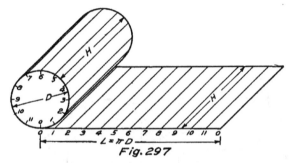

Fig. 297

distance around the prism. At each of the points a line is drawn equal to the long edge of the prism and the ends joined together. Then the two ends of the prism are measured out as shown.

The development of the lateral surface of a hexagonal prism

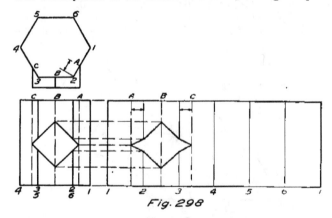

Fig. 298

is shown in Fig. 298. First lay off in a straight line and in proper order the edges *1–2*, *2–3*, etc., all the way around the prism as shown at the right. At points *1*, *2*, *3*, etc., draw the perpendiculars equal in length to the edges of the prism, thus obtaining the true size and shape of each face of the prism and in such order that they might be folded to the form of the prism. Note that a square prism intersects the hexagonal prism which has been cut along the curve of intersection. To find the cut-out on the de-

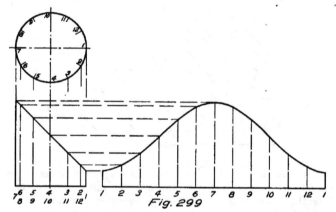

Fig. 299

velopment draw the vertical lines A, B, and C on the faces of the hexagonal prism and locate them on the development by measuring their distances from the edges *2* and *3*. The points of inter-

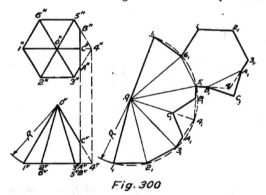

Fig. 300

section may then be located by drawing horizontal lines as shown or by measuring up or down the lines on the front view of the hexagonal prism and measuring the same distances on the same lines of the development. The development of the top and bottom of the prism may be obtained from the top view and added to the lateral surface.

Development of a Cylinder. — The development of a cylinder was illustrated in Fig. 297. One half of a square elbow is developed in Fig. 299. First divide the top view into a number of equal parts. Through each point draw an elemen of the cylinder. By taking the elements close enough together the arcs may be considered as straight lines. The problem is then the same as developing a prism with a large number of sides. Lay off the distances between the elements along a straight line. At each point draw the element in its true length. Through

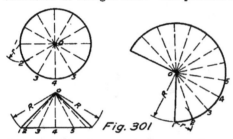

Fig. 301

the ends of the elements draw a smooth curve, very lightly freehand, and then brighten it up using the irregular curve. The lengths of the elements may be conveniently found by drawing horizontal lines from the front view as illustrated. The development of the bases may be found by an auxiliary view and from the top view.

Development of a Pyramid. — The development of a pyramid with a part cut away is shown in Fig. 300. Assume the pyramid to be complete. There are six equal faces, each one a triangle. The development consists in laying out all the faces in their true size and proper order. The short edges are shown in their true length in the top view as $1–2$, $2–3$, etc. The long edges are all of the same length and are equal to the distance $O^V–1^V$ shown in the front view. Observe that $O^H–1^H$ is horizontal in the top view. The faces may be constructed in their true size by drawing an arc, with O_1 as a center and O^V1^V as a radius. Starting at 1_1, space off the chords $1_1–2_1$, $2_1–3_1$, etc., equal to $1^H–2^H$, $2^H–3^H$, etc. Draw lines O_1 1_1, O_1 2_1, etc., representing edges of the pyramid. Construct the development of the base so that it may be folded into the proper position. Note carefully that the numbers on the base will match the numbers on the edges when the development is folded to form the pyramid. To show the part which

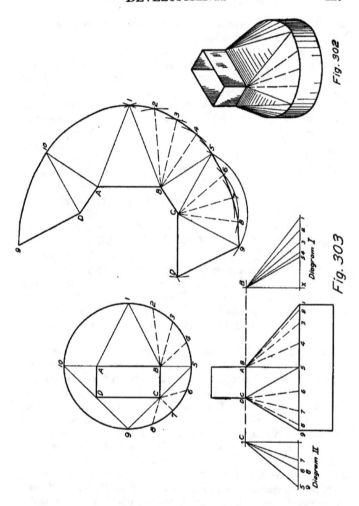

Fig. 302

Fig. 303

has been cut away measure the distance O_1C_1 on the edge O_1A_1 equal to the distance O^vC^v. Measure distances 5_1B_1 and 3_1A_1 on the development of the faces and of the base equal to the distances 5^HB^H and $3^H A^H$ obtained from the top view. Join points A_1C_1

and B_1 on the development of the faces. On the development of the base construct the triangle $A_1B_1C_1$ obtaining distances B_1C_1 and A_1C_1 from the development of the faces. The completed development is shown by the heavy lines.

The Development of a Cone. — The development of a cone is shown in Fig. 301. Divide the base into a number of parts and draw elements of the cone. By taking the small arcs as straight lines the solution is the same as for a pyramid. The surface is thus considered to be divided into a number of equal triangles. This method is sufficiently accurate for most purposes. With the radius R draw an arc of a circle. On the arc space off the circumference of the base of the cone. The base need not be developed as it shows in its true size in the top view.

Development of a Transition Piece. — A transition piece is shown in Fig. 302 connecting a circular pipe with a rectangular one. The development of such a piece should present no difficulties if the previous figures have been carefully studied. Comparing the two views as given in Fig. 303 with the picture of Fig. 302, it will be seen that the transition piece may be "broken up" into triangles and parts of cones. The triangles are $AB1$, $BC5$, $CD9$, and $DA10$, The parts of cones are the curved surfaces between the triangles. Consider the apex of one cone as located at B. Divide the portion of the base 1–5 into a number of parts and draw the elements B–1, B–2, B–3, B–4, and B–5. The triangles thus formed will approximate the surface of the cone. The lines AB, BC, etc. show in their true length in the top view. The true length of the elements may be found as follows: Consider a line to be dropped from point B perpendicular to the base of the cone. A line may then be drawn on the base of the cone from point 1 to the perpendicular line, thus forming a right triangle with the element B–1 as the hypotenuse. By constructing this right triangle in its true size the true length of B–1 may be found. This has been done in Diagram I. The length of the perpendicular line is shown at BX and is found by drawing the horizontal lines shown. The base of the triangle is equal to the length of the horizontal projection of B–1. Point 1 in Diagram I is found by making x–1 equal to B–1. In the same way find the lengths of the other elements by laying off

$$x\text{–}2 \text{ equal to } B\text{–}2$$
$$x\text{–}3 \text{ equal to } B\text{–}3$$

etc., obtained from the top view. Then draw *B–2*, *B–3*, etc. the true lengths of the elements which are used in the construction described below. In the same manner construct Diagram II for the other cone. Having found all the true lengths proceed as follows: Construct the triangle *AB1*, in its true size. With *B* as a center and *B2* as a radius, draw an arc. With *1* as a center and a radius equal to *1–2* obtained from the top view describe another arc cutting the first arc. This will locate point *2*. With *B* as a center and *B–3* as a radius describe an arc. With *2* as a center and a radius equal to *2–3* obtained from the top view describe another arc, thus locating point *3*. Proceed until the four triangles forming the conical surface are properly located, then draw a smooth curve through the points *1*, *2*, *3*, etc. Construct triangle *CB5*, using the element *B5* as a starting side. Then develop the conical surface having *C* as an apex and *5*, *6*, *7*, *8*, *9*, as part of the base. Construct the triangle *CD9* in its true size. Since the piece is symmetrical the remaining parts are the same as those already developed.

All kinds of surfaces can be developed approximately by dividing them into triangles, then finding the true size of each triangle and arranging them in the proper relation to each other.

CHAPTER XVIII

PICTURE DRAWING

Isometric Drawing. — By means of an isometric projection three faces of an object can be shown in a single view. This is possible by considering the object to be placed in the position of a cube standing on one corner and having another corner exactly

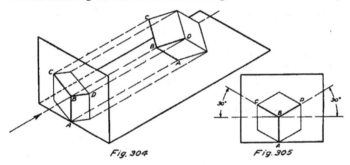

Fig. 304   Fig. 305

in the center of the view. In Fig. 304 the cube is resting upon point A in such a position that point B is located in the center of

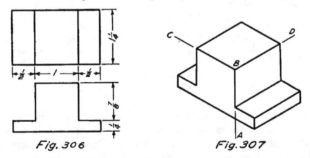

Fig. 306   Fig. 307

the view obtained by projecting onto a vertical plane. The orthographic projection of this front view is shown in Fig. 305, which is called the isometric projection of a cube. In this view

130

the line AB is vertical and the lines BC and BD make angles of 30° with the horizontal. All the edges of the cube show equal to each other in length. This length however is shorter than

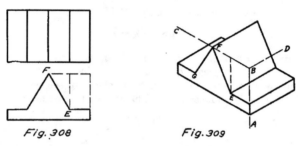

Fig. 308   Fig. 309

on the actual cube. For drawing purposes the lines BD, BC, and BA, etc. are made the same length as on the actual cube. The angles formed by the three lines which meet at point B are equal to 120° each. The three lines are called the isometric axes and form the basis for isometric drawing.

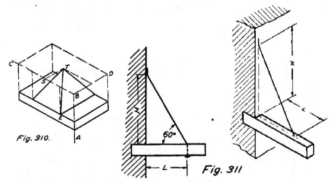

Fig. 310.   Fig. 311

Isometric and Non-isometric Lines. — All measurements for isometric drawings are taken along or parallel to the isometric axes. Lines parallel to the isometric axes are called isometric lines. All other lines are non-isometric lines and cannot be measured directly.

To make an Isometric Drawing of the Object shown in Fig. 306. — Draw the isometric axes, BC, BA, and BD (Fig. 307).

From B measure $1^{1}/_{4}''$ toward D, $1''$ toward C, and $^{7}/_{8}''$ toward A. From the points thus located draw lines parallel to the isometric axes and lay off distances corresponding to the figures given in

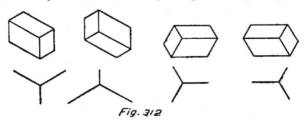

Fig. 312

Fig. 306. Note that lines which are parallel in Fig. 306 are parallel in Fig. 307.

To make an Isometric Drawing of the Object shown in Fig. 308. — Draw the isometric axes (Fig. 309) as in the preceding case. Locate the point F by measuring along BC. Locate point E by measuring along BC and then down parallel to BA

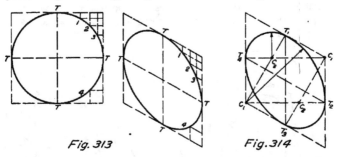

Fig. 313   Fig. 314

as indicated in the figure. Join F and E. Line FE is a non-isometric line.

In Fig. 310 point E is located as before. Point T is located by measuring along BC to point S and then parallel to line BD. It is often convenient to think of the object as being placed in a box. This box can be put into isometric and the points in which the object touches it located. Other points can be located by taking measurements parallel to the axes.

Angles. — Angles do not show in their true size in isometric drawings. This is evident from an inspection of Fig. 305 where the angle at B is 120° and that at C is 60° although on the cube they are both 90°. The method of constructing for angles is shown in Fig. 311. First make the orthographic projection, then transfer by taking distances parallel to the axes, as H and L.

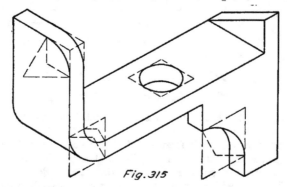

Fig. 315

Positions of the Axes. — The axes may be placed in any position provided the angles between them are kept equal to 120° as illustrated in Fig. 312.

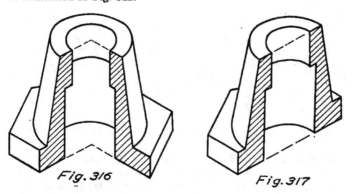

Fig. 316   Fig. 317

Construction for Circles. — When circles occur they appear as ellipses and may be drawn by plotting points from the orthographic projection as in Fig. 313 or by the more usual approximation shown in Fig. 314, where the lines are drawn perpendicular to the points of tangency of the circumscribing square. The intersections of these perpendiculars locate the centers for circular arcs which will approximate the ellipse sufficiently close for most purposes. In the figure

$T_1 T_2 T_3 T_4$ = *tangent points*
C_1 = *center for arc* $T_1 T_2$ *and* $T_3 T_4$
$C_1 T_1$ = *radius for arc* $T_1 T_2$ *and* $T_3 T_4$
C_2 = *center for arc* $T_1 T_4$ *and* $T_2 T_3$
$C_2 T_1$ = *radius for arc* $T_1 T_4$ *and* $T_2 T_3$

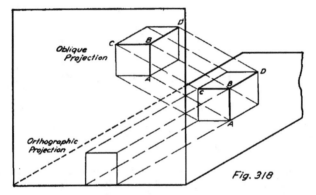

Fig. 318

The same construction is used for arcs of circles as shown in Fig. 315.

The interior of objects may be shown by means of isometric sectional views, Fig. 316 and Fig. 317, which are constructed by

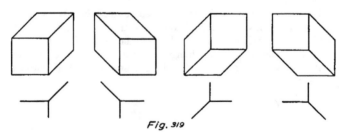

Fig. 319

the methods already described for exterior views. As shown, the sectioned surfaces are taken on isometric planes.

Oblique Drawing. — Another method of picture drawing often useful is oblique drawing or projection, in which the view is obtained by using projection lines oblique to the plane upon which the object is to be represented. In Fig. 318 the orthographic projection of a cube is shown and, on the same plane, the oblique projection of the same cube. The three lines which meet at point *B* are called oblique axes. Lines *BC* and *AB* are always at right angles but the line *BD* may make any convenient angle

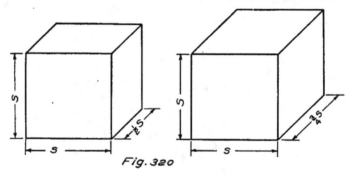

Fig. 320

with the horizontal. It follows that if one face of an object is parallel to the vertical plane, it will show in its true size and shape.

After locating the axes the methods of construction given for isometric drawing apply to the making of oblique drawings. Many examples of oblique drawing are given throughout this book. The axes may be located in a variety of ways as shown in Fig. 319.

The appearance of an object can often be improved by reducing the measurements along the oblique axis, using one half or three fourths of the full dimension. Measurements on the two perpendicular axes remain unchanged. Two such treatments of a cube are shown in Fig. 320. Such views are called cabinet projection.

CHAPTER XIX

SHADE LINE DRAWINGS

Shade Lines. — The use of shade lines is a much discussed question. Each drawing has a purpose and if that purpose is better served by the use of shade lines they should be employed.

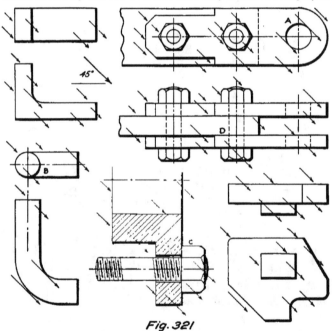

Fig. 321

In many lines of work detail drawings are never shaded and this seems to be the best practice. Outline drawings or assembly drawings which serve partly, at least, as picture drawings are often improved by shading.

136

System in Common Use. — In the United States a conventional system of shading is generally employed, in which the rays of light are assumed to be parallel, to come from the upper

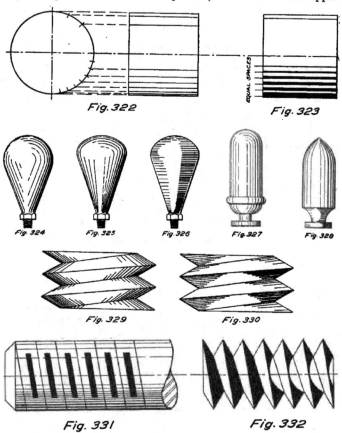

Fig. 322   Fig. 323

Fig. 324   Fig. 325   Fig. 326   Fig. 327   Fig. 328

Fig. 329   Fig. 330

Fig. 331   Fig. 332

and left hand corner of the sheet at an angle of 45°, and to lie in the plane of the paper. The lower and right hand edges where the light passes over them are made heavy lines called shade lines. When two surfaces are in the same plane the line of division between them is not shaded, Fig. 321. Circles follow the same

rules as shown, where A is a hole and B is a solid cylinder.   In
all cases the extra thickness of line is without the surface which

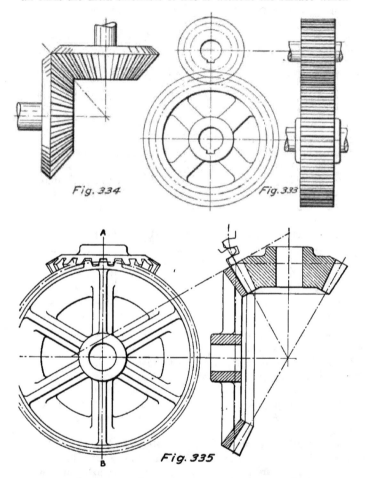

Fig. 334

Fig. 333

Fig. 335

it bounds (C, Fig. 321).   Most all conditions of shading are
illustrated in the figures given in this chapter, which should
be carefully studied.

Surface Shading. — Various methods of line shading on surfaces are used to show the shape of machine parts. Personal judgment is an important element in the matter of successful surface shading. Fig. 322 shows a cylinder shaded by using fine lines and varying the distances between them. These change approximately as the projections of equally spaced elements of a cylinder.

Fig. 336

Another method is to space the shade lines about equally but to vary the width of the lines as in Fig. 323. The air chambers (Figs. 324 to 328) show a number of different ways of shading conical, spherical, and cylindrical surfaces. As shown, either fine lines near together or varying lines may be used with any of the methods illustrated.

Shading Screw Threads and Gears. — On elaborate drawings

Fig. 337   *Fig. 338*

it is sometimes desirable to shade screw threads. Five ways are shown in Figs. 329 to 332.

When gears are to be shown without drawing in the teeth the exterior is frequently represented by alternating heavy and fine lines as in Figs. 333 and 334 which show a pair of bevel gears and a pair of spur gears. The rest of the drawing may or may not be shaded. A pair of bevel gears are shown in section in Fig. 335.

Special Surface Representations. — Other surfaces may be represented as in Figs. 336 to 338. Three ways of indicating a knurled surface are given in Fig. 336. For a scraped surface Fig. 337 may be used, and for a polished surface, Fig. 338.

Patent Office Drawing. — Probably the most general use of shaded drawings is for Patent Office work. Such drawings must

be made on pure white paper of a thickness equal to two or three
ply Bristol board, using black ink. The outside dimensions of
the sheet are 10 by 15 inches. Inside of this is a one inch margin.
At the top of each sheet a clear space of one and one quarter
inches must be left for a title which is printed in by the Patent

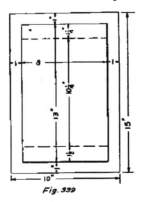

Fig. 339

Office. Fig. 339 shows the layout of a patent drawing. The
fewest number of lines should be used; all dimension and center
lines should be left off. The plane upon which a section is taken
should be indicated. All parts are lettered or numbered. As
these drawings are reproduced by the photo zinc process, all lines
must be absolutely black and not too fine. If lines are too close
together they will run together when printed. The "Rules of
Practice" of the United States Patent Office may be had for the
asking and should be consulted by those interested.

CHAPTER XX

DRAWING QUESTIONS, PROBLEMS, AND STUDIES

Most of the drawing studies included in this chapter can be worked in an 11″ × 14″ space or in a division of the space as indicated in Figs. 340 to 343. The layout with dimensions for a regular size sheet is shown in Fig. 340. In some cases a large scale may be advisable in which case the full sheet may be used. An inspection of the problem will indicate the proper space where it is not given in connection with the problem. The order in

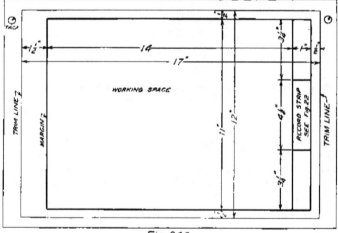

Fig. 340

which the problems are given can be varied to suit the needs of the class. The question of inking is left for the instructor to decide. The author advises that it be delayed until the student has attained considerable proficiency in making pencil drawings. A variety of problems is included to allow a selection to be made and so that the course may be varied from year to year. A number of answers to questions should be neatly written or

141

lettered and numerical problems should be carefully worked out to create a coördination between drawing and other subjects, as well as to impress the student with the fact that the mere drawing of lines is not the aim of a drawing course. It is thought

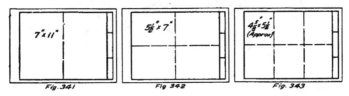

Fig. 341   Fig 342   Fig. 343

that such problems may create an interest and stir the student with the ambition to seek an engineering education.

1. Describe the proper use of the T square.

2. Show by a sketch the proper method of sharpening a lead pencil.

3. How are horizontal lines drawn?

4. How are vertical lines drawn?

5. Show by sketches the proper adjustment of the pen, pencil, and needle points for a compass.

6. Draw a straight line. Draw short lines crossing this line, and $2^3/_{16}''$ apart. Draw another short line crossing the original line, and $1^9/_{16}''$ from the last line drawn. From this lay off further distances of $1^7/_8''$ and $^{15}/_{16}''$. Add the four distances and check the total length by scaling the line. In measuring a line, place the zero of the scale opposite one end of the line and read the scale opposite the other end of the line.

7. Draw a straight line. Set the dividers at $^9/_{16}''$ and step off 10 spaces. Scale the distance thus found and check with the calculated length.

8. What is the purpose of the knee joints in the compasses?

9. Examine a drawing material catalog and list five tools in addition to those which you already have, that you would consider convenient for your work.

10. What kinds of pens are used for freehand lettering?

11. What kind of ink is used?

12. What is the slope for slant letters?

13. In what direction should the pen point?

14. How is the amount of ink on the pen regulated?

15. What hardness of pencil should be used for lettering?

16. How is the distance between letters regulated?

17. $11'' \times 14''$ space. Starting $1/2''$ from top border line draw horizontal guide lines $3/8''$ apart. Use very light pencil lines. Make each capital letter of Fig. 15 or 16 five times. Repeat the letters which cause most trouble. Use *2H* pencil.

18. $5^1/2'' \times 7''$ space. Starting $1/2''$ from top border line draw horizontal guide lines $1/8''$ apart. Make each of the lower case letters of Fig. 15 or 16 five times. Height of letters *a, c, e,* etc. to be $1/4''$. Height of letter *b, k,* etc. to be $3/8''$. Use *2H* pencil.

19. $5^1/2'' \times 7''$ space. Same as problem 18, but use ball pointed pen.

20. $5^1/2'' \times 7''$ space. Starting $1/2''$ from top border line draw horizontal guide lines $1/4''$ apart. Make each capital letter of Fig. 15 or 16 five times. Use $2H$ pencil.

21. $5^1/2'' \times 7''$ space. Same as problem *20*, but use ball pointed pen.

22. $5^1/2'' \times 7''$ space. Starting $1/2''$ from top of space draw horizontal guide lines $1/8''$ apart. Letter the following words, using a *2H* pencil: HILL, LATE, LATHE, BOLT, QUENCH, WRENCH, EQUIPMENT, TOOLS, CALIPERS

23. $5^1/2'' \times 7$ space. Same as problem 22, but using pen and ink as directed.

24. $5^1/2'' \times 7''$ space. Draw horizontal guide lines near the middle of the space for letters having $1/8''$ caps. Letter the following, using a *2H* pencil. Use caps and lower case of Fig. 15 or 16.

"Drawing is the education of the eye, it is more interesting than words. It is the graphic language."

"Mechanical drawing is the alphabet of the engineer; without this the workman is merely a hand, with it he indicates the possession of a head."

25. Prepare a title and material list for the step bearing shown in Fig. 175.

26. Same as problem 24, but using pen and ink as directed by the instructor.

27. Name and illustrate three kinds of triangles.

28. Name and illustrate three kinds of quadrilaterals.

29. What is a right angle?

30. In order that the sills of a house may be square 6 feet has been measured off along one sill and 8 feet along the other. Nails are driven as in Fig. 344 at these points. What will be the distance AC measured along a steel tape when the angle ABC is a right angle?

31. A circle has a diameter of 2 inches. What is its circumference? Compare this distance with the sum of the sides of an inscribed hexagon.

32. What is an ellipse?

33. Can a true ellipse be drawn with circular arcs?

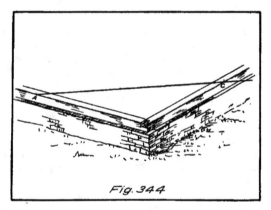

Fig. 344

34. Space $4^5/8''$ wide, $5^1/2''$ high. Draw a line $2^{15}/_{16}''$ long and bisect it. See Fig. 32.

35. Space as for problem 34. Draw an angle and bisect it. See Fig. 33.

36. Space as for problem 34. Draw a line $2^{13}/_{16}''$ long and divide it into five equal parts, by method of Fig. 34.

37. Space as for problem 34. Same as problem 36, but use method of Fig. 35.

38. Space as for problem 34. Draw any angle and construct another angle equal to it. See Fig. 36.

39. Space as for problem 34. Construct a triangle, having sides as follows: $2^5/8''$; $3^1/8''$; and $2''$. See Fig. 37.

40. Space as for problem 34. Construct an equilateral triangle, one side $2^9/_{16}''$ long. See Fig. 38.

41. Space as for problem 34. Draw an isosceles triangle having a base of $2^7/_8''$. Sides make 75° with the base. See Fig. 28.

42. Space as for problem 34. Draw a right triangle. Hypotenuse $3^1/_2''$ long. One angle is 30°.

43. Space as for problem 34. Mark three points (+) not in a straight line, and draw a circle passing through them. See Fig. 42.

44. Space as for problem 34. Draw an arc of a circle. Radius $2''$, with center $^1/_2''$ from upper and left hand edges of space. Make the angle AOB (Fig. 43) equal to 45°. Find length of the arc. Use first method of Fig. 43.

45. Space as for problem 34. Same as problem 44, but use second method of Fig. 43.

46. Space as for problem 34. Draw a circle $2^5/_8''$ diameter. Draw a tangent at any point on the circumference. See Fig. 44.

47. Space as for problem 34. Draw an arc with a radius of $1^1/_4''$. Draw a straight line intersecting this arc. Draw an arc tangent to the arc and straight line just drawn, radius $^5/_8''$. See Fig. 45.

48. Space as for problem 34. Draw a hexagon in a circle having a diameter of $2^7/_8''$. See Fig. 39.

49. Space as for problem 34. Draw a hexagon having a measurement across flats (Fig. 39) of $2^1/_4''$.

50. Space as for. problem 34. Draw a regular octagon inside of a $3^1/_8''$ square, See Fig. 40.

51. Space as for problem 34. Draw a regular octagon inside of a $3^1/_8''$ circle.

52. Space as for problem 34. Draw a right triangle having a hypotenuse $3''$ long, and one side $2''$ long. Draw a circle passing through the points of the triangle.

53. Space as for problem 34. Draw a line $3^1/_8''$ long and divide it into parts proportional to 2, 3, and 5. Use a method similar to Fig. 35.

54. Space as for problem 34. Use 30° × 60° triangle and T square to draw a regular hexagon measuring $3^7/_{16}''$ across corners.

55. Space as for problem 34. Using the 45° triangle and T square draw a regular octagon that will just contain a circle $3^1/_{16}''$ diameter.

56. Space $5^1/_2'' \times 7''$. Draw an ellipse by concentric circle method (Fig. 49). Major axis 5''. Minor axis 2''. Find 24 points.

57. Space $5^1/_2'' \times 7''$. Draw an ellipse by trammel method (Fig. 50). Major axis $5^1/_4''$. Minor axis $2^1/_8''$.

58. Space $5^1/_2'' \times 7''$. Draw a figure having the appearance of an ellipse by circular arcs, Fig. 51. $AB = 5''$, $CD = 2^3/_4''$.

59. Space $5^1/_2'' \times 7''$. Draw a $^3/_4''$ square in the center of the space. Draw an involute of this square.

60. Space $5^1/_2'' \times 7''$. Draw a semi-circle having its center $^1/_4''$ from the left edge of the space and $3^1/_4''$ down from the top

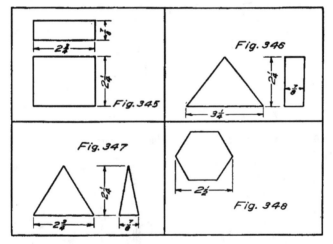

Fig. 345

Fig. 346

Fig. 347

Fig. 348

of the space. Radius of circle $1^3/_4''$. Draw the involute of the semi-circle. See Fig. 53.

61. Space $5^1/_2'' \times 7''$. Draw a parabola, Fig. 54. Distance $AF = ^7/_8''$. Directrix perpendicular.

62. Space $5^1/_2'' \times 7''$. Draw a parabola, Fig. 54. Directrix horizontal. Distance $AF = ^1/_4''$.

63. Space $5^1/_2'' \times 7''$. Draw an equilateral hyperbola, Fig. 56. Point P is $1''$ from line OG and $2^1/_8''$ from OH. Make distances PA, AB, etc. $^1/_4''$.

64. Space $5^1/_2'' \times 7''$. Draw the two views given and the side view of the prism, Fig. 345.

65. Space $5^1/_2'' \times 7''$. Draw the two views given and the top view of Fig. 346.

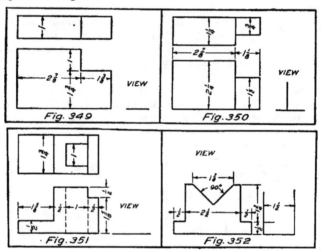

Fig. 349   Fig. 350   Fig. 351   Fig. 352

66. Space $5^1/_2'' \times 7''$. Draw the two views given and the top view of Fig. 347.

67. Space $5^1/_2'' \times 7''$. Draw the top, front, and side views of a regular hexagonal prism, Fig. 348. Corners of hex $2^1/_2''$. Height of prism $1/_2''$.

68. Space $5^1/_2'' \times 7''$. Draw three views of the object shown in Fig. 349.

69. Space $5^1/_2'' \times 7''$. Draw three views of the object shown in Fig. 350.

70. Space $5^1/_2'' \times 7''$. Draw three views of the object shown in Fig. 351.

71. Space $5^1/_2'' \times 7''$. Draw three views of the object shown in Fig. 352.

72. Space $5^1/_2'' \times 7''$. Draw three views of the object shown in Fig. 353.

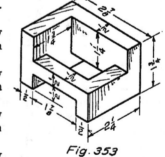

Fig. 353

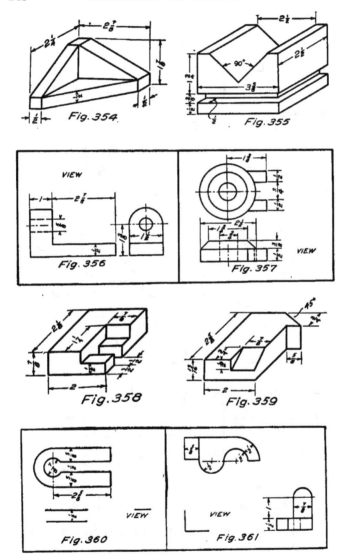

Fig. 354.

Fig. 355

Fig. 356

Fig. 357

Fig. 358

Fig. 359

Fig. 360

Fig. 361

73. Space $5^1/_2'' \times 7''$. Draw three views of the object shown in Fig. 354.

74. Space $5^1/_2'' \times 7''$. Draw three views of the object shown in Fig. 355.

75. Space $5^1/_2'' \times 7''$. Draw three views of the object shown in Fig. 356.

76. Space $5^1/_2'' \times 7''$. Draw three views of the object shown in Fig. 357.

77. Space $5^1/_2'' \times 7''$. Draw three views of the object shown in Fig. 358.

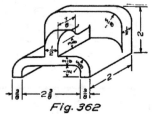

Fig. 362

78. Space $5^1/_2'' \times 7''$. Draw three views of the object shown in Fig. 359.

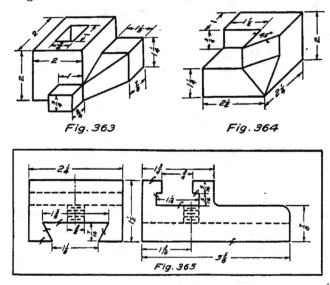

Fig. 363   Fig. 364

Fig. 365

79. Space $5^1/_2'' \times 7''$. Draw three views of the object shown in Fig. 360.

80. Space $5^1/_2'' \times 7''$. Draw three views of the object shown in Fig. 361.

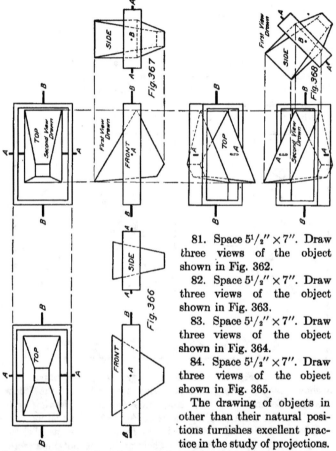

81. Space $5^1/_2'' \times 7''$. Draw three views of the object shown in Fig. 362.

82. Space $5^1/_2'' \times 7''$. Draw three views of the object shown in Fig. 363.

83. Space $5^1/_2'' \times 7''$. Draw three views of the object shown in Fig. 364.

84. Space $5^1/_2'' \times 7''$. Draw three views of the object shown in Fig. 365.

The drawing of objects in other than their natural positions furnishes excellent practice in the study of projections. It is a useful test of one's knowledge of the theory of drawing, and every student should have some experience with such problems.

The method of solution for such problems calls for the location of each point in its three views and particular attention to relations of the lines.

Three views of a hopper are shown in Fig. 366. When the hopper is revolved to 30° about shaft AA, the front view will show as in Fig. 367. The top view is obtained by projecting horizontally from the top view of Fig. 366 and vertically from the front view of Fig. 367. The front view of Fig. 367 is changed only in the position of the hopper. In the top view the distances parallel to the shaft *AA* have not been changed, as the revolution has been about this axis. The side view of Fig. 367 is then obtained in the usual manner from the top and front views.

With the apparatus in the position of Fig. 367, it may be revolved about the shaft *BB* forward or backward. In this case

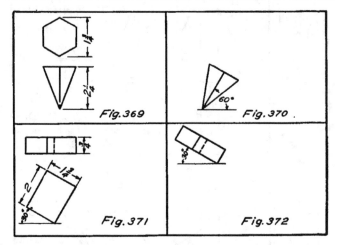

Fig. 369

Fig. 370

Fig. 371

Fig. 372

the side view of Fig. 367 will be unchanged except for its position as shown in Fig. 368. After drawing the side view the front view may be drawn by projecting across from the side view and down from the front and top views of Fig. 367. This is possible because the horizontal distances in the front view are parallel to the shaft or axis of revolution. The top view is obtained from the other two views in the usual way.

85. Space $5^1/_2'' \times 7''$. Draw three views of the hexagonal pyramid in the position shown in Fig. 369.

86. Space $5^1/_2'' \times 7''$. Draw three views of the pyramid of Problem 85 after it has been revolved as shown in Fig. 370.

87. Space $5^1/_2'' \times 7''$. Draw three views of the rectangular prism in the position shown in Fig. 371.

88. Space $5^1/_2'' \times 7''$. Draw three views of the rectangular prism after it has been revolved from the position of Fig. 371 about a vertical axis. Top view is shown in Fig. 372.

89. Space $5^1/_2'' \times 7''$. Draw three views and a complete auxiliary view of the square prism shown in Fig. 373, after it has been cut by plane A–A and the part above the plane removed.

90. Space $5^1/_2'' \times 7''$. Draw two views and a complete auxiliary view of the hexagonal prism shown in Fig. 374, after it has been cut by plane A–A.

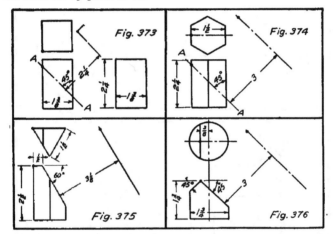

91. Space $5^1/_2'' \times 7''$. Draw the two views given and a complete auxiliary view, Fig. 375.

92. Space $5^1/_2'' \times 7''$. Draw the two views given and a complete auxiliary view, Fig. 376.

93. Space $5^1/_2'' \times 7''$. Draw a complete auxiliary view, Fig. 377.

94. Space $5^1/_2'' \times 7''$. Draw a complete auxiliary view, Fig. 378.

95. Space $11'' \times 14''$. Draw the two views shown and an auxiliary view of the foot pedal shown in Fig. 379.

96. Space $11'' \times 14''$. Complete the views and draw an auxiliary view of the molding, Fig. 380.

97. Why are sectional views used?

98. What is the relation of a sectional view to the other views?

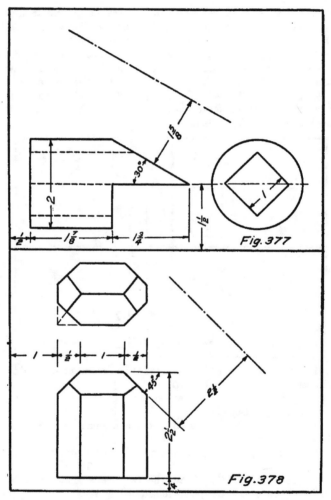

Fig. 377

Fig. 378

99. Space $5^{1}/_{2}'' \times 7''$. Draw a sectional view of Fig. 381 on a plane through the axis.

100. Space $5^{1}/_{2}'' \times 7''$. Draw a sectional view of Fig. 382 on a plane through the axis.

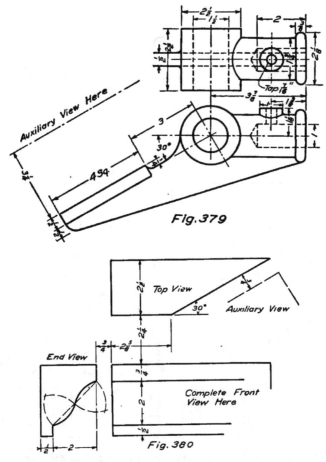

Fig. 379

Fig. 380

101. Space $5^1/_2'' \times 7''$. Draw a sectional view of Fig. 383 on a plane through the axis.

102. Space $5^1/_2'' \times 7''$. Draw three views of Fig. 384, changing the proper view to a section on plane $A–A$.

103. Space $5\frac{1}{2}'' \times 7''$. Draw two views of Fig. 385, changing the proper view to a section on plane $A–A$.

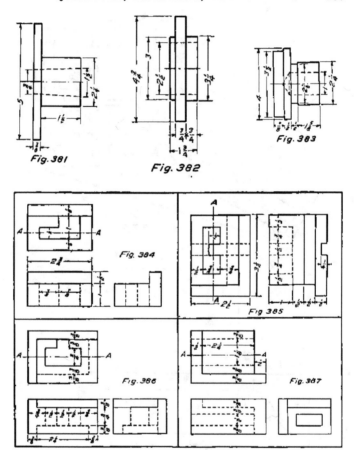

Fig. 381

Fig. 382

Fig. 383

Fig. 384

Fig 385

Fig. 386

Fig. 387

104. Space $5^{1}/_{2}'' \times 7''$. Draw three views of Fig. 386, changing the proper view to a section on plane $A-A$.

105. Space $5^{1}/_{2}'' \times 7''$. Draw three views of Fig. 387, changing the proper view to a section on plane $A-A$.

106. Space $11'' \times 14''$. Draw three views of the slide valve, Fig. 388. The missing view to be a section on plane $A-A$.

107. Draw three views of the shackle, Fig. 186.

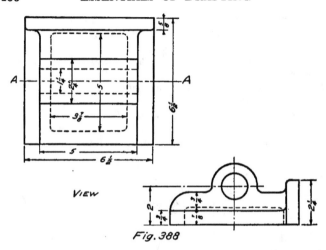

VIEW

Fig. 388

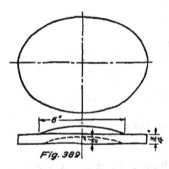

Fig. 389

108. Space 11" × 14". Draw a plan view and a sectional elevation for the elliptical cover plate shown in Fig. 389. Outside dimensions 7" × 9". Six ¹¹/₁₆ inch holes for bolts. The rise in the center is elliptical in plan. The bolts are to be spaced equal distances apart. Draw full size.

109. 11" × 14" space. Draw two views of the crank shown in Fig. 390. This drawing is excellent as an inking or tracing exercise.

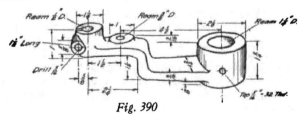

Fig. 390

110. Compare briefly wrought iron and cast iron.

111. What is cast iron? Name some of its properties. Compare its strength in tension and compression.

112. What is wrought iron? How is it made? Name some of its properties.

113. What is steel? How is it made? Name some of its properties.

114. What material is used for bolts and nuts?

115. How is malleable iron made and what is it used for?

116. Of what material would you make the following and why?

> a. Steam Engine Cylinder.
> b. Water Pump Plunger.
> c. Piston-rod.
> d. Complicated form of Lever.
> e. Shaft.

117. What is meant by unit stress? Axial stress? Compression? Tension? Shear?

118. A tie-bar has a diameter of $7/8''$ and supports a load of 8000 pounds. What is the unit stress?

119. What load will a rectangular tension member measuring $3/8'' \times 1''$ carry safely if it is made of wrought iron? (Live load.)

120. A hollow cast iron cylinder has diameters of $4''$ and $3''$. What safe compressive load will it carry if the load is steady?

121. Compute the number of $3/4''$ bolts for a cylinder head $15''$ effective diameter. Steam pressure is 150 pounds per square inch. Allowable working stress on bolts is 5000 pounds per square inch. The effective root area of a $3/4''$ bolt is .302 square inches.

122. A wrought iron bolt $1 1/2''$ diameter has a head $1 1/4''$ long. Its effective diameter is 1.284. When a tension of 14000 pounds is applied to the bolt, find the unit stress.

123. What are some of the uses of screw threads?

124. What advantage has the acme thread over the square thread?

125. A triple threaded screw has a pitch of $1/8$ inch. How many turns must it make to move a nut 6 inches?

126. Express the following in terms of the diameter of the bolt; distance across flats of hex, thickness of bolt head, and thickness of nut.

127. In what way does a bolt head differ from a nut?

128. Draw a 1" hex nut across flats and a 1" square nut across corners. Compare them.

129. Space $5\frac{1}{2}$" × 7". Draw one turn each of two helices as started in Fig. 391.

130. Space $5\frac{1}{2}$" × 7". Draw the exterior of a square threaded screw 3" long which enters 1" into the section of a square threaded nut. Pitch 1". Other dimensions as in Fig. 392.

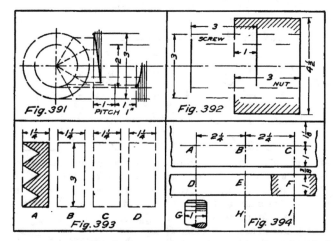

131. Space $5\frac{1}{2}$" × 7". Draw four forms of screw threads in section as directed by the instructor. 1" pitch. Fig. 393.

132. Space $5\frac{1}{2}$" × 7". Fig. 394. At A, B, and C, draw three different plan views of threaded holes. At D and E draw two different representations of threaded holes in elevation. At F draw a threaded hole in section. At G, H, and I, draw three conventional representations of threaded bolt ends. Diameter for all representations to be 1 inch.

133. Space $5\frac{1}{2}$" × 7". On axis A–B, Fig. 395, draw a $\frac{3}{4}$" through bolt, hex head across corners and hex nut across flats. On axis C–D draw a $1\frac{1}{8}$" bolt, hex head across flats and hex nut across corners. Indicate required dimensions.

134. Space $5\frac{1}{2}$" × 7". On axis A–B, Fig. 396, draw a $\frac{7}{8}$" bolt, square head across corners and square nut across flats.

On axis $C-D$ draw a $^7/_8''$ cap screw, head across flats.   On axis $E-F$ draw a $^7/_8''$ cap screw, head across corners.

135. Space $5^1/_2'' \times 7''$.   Draw the two views of collar and

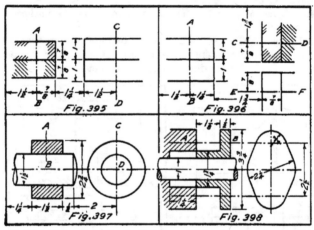

Fig. 395   Fig. 396   Fig. 397   Fig. 398

shaft, Fig. 397.   On axis $A-B$ draw a $^5/_8''$ set screw, head across flats.   On axis $C-D$ draw same set screw, head across corners.

136. Space $5^1/_2'' \times 7''$.   Draw the gland and stuffing box of Fig. 398.   On axis $A-B$ draw a $^1/_2''$ stud and nut.   Show nut across flats.   Make provision for the gland to enter one half the depth of the stuffing box when nut is screwed onto stud. Show required dimensions.

137. Draw a plan and section for a double riveted lap joint as directed by the instructor.

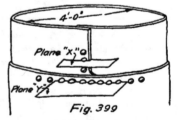

Fig. 399

138. Make a scale drawing of two plates joined together at right angles.

139. Space $5^1/_2'' \times 7''$.   Draw sections on planes X and Y and a development of the riveted joint of Fig. 399.   See Chapter VIII. $^7/_{16}''$ plates; $^{15}/_{16}''$ rivets; pitch $2^7/_{16}''$; lap $1^1/_2''$; scale $3'' = 1$ foot.

140. How many views should a drawing contain?

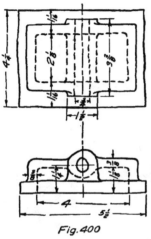

Fig. 400

141. What scales are in general use for working drawings?

142. What are the first lines inked on a working drawing?

143. Is true projection always used? Explain.

144. Sketch and describe one form of stuffing box.

145. Space 11″ × 14″. Make a working drawing showing three views of the slide valve shown in Fig. 400. One view may be a section. Completely dimension.

146. Space 11″ × 14″. Make a working drawing of the bearing cap of Fig. 401. Show three views. Completely dimension. One view may be a section. Supply any missing dimensions. See Chapter XV for size of ¹/₄″ pipe.

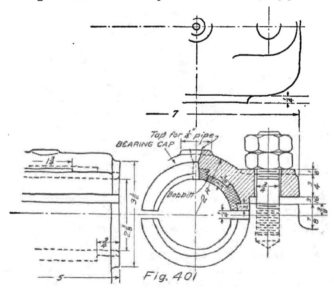

Fig. 401

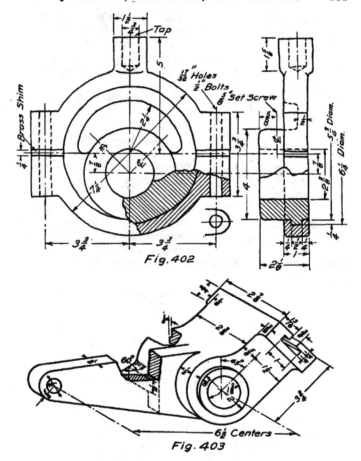

Fig. 402

Fig. 403

147. Make detail working drawings for the parts of the eccentric shown in Fig. 402. Supply any missing dimensions. Drawing should include a properly dimensioned bolt and set screw. Completely dimension the drawing.

148. Make detail working drawings for the parts of the step bearing shown in Fig. 175. Scale 6″ = 1 foot. Use two sheets, 11″ × 14″ space. Completely dimension.

149. Draw two views of Fig. 403. Each view should show true distances. Completely dimension. Submit a preliminary sketch to the instructor.

150. Make a working drawing for the piece shown in Fig. 404. Submit a preliminary sketch to the instructor.

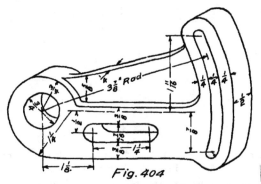

Fig. 404

151. Space $5^1/_2'' \times 7''$. Make a detail working drawing of the construction shown in Fig. 405, using one full view and such parts of other views as are necessary to define its true shape.

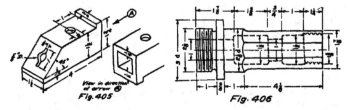

Fig. 405

Fig. 406

152. Space $5^1/_2'' \times 7''$. Make a working drawing of the sleeve, Fig. 406. One half view to be in section.

153. Make a working drawing of the valve shown in Fig. 407. Show a proper treatment for a section on plane ABC.

154. Draw a sectional view of Fig. 408.

155. Make a detail drawing of the valve body of Fig. 409. One view in section.

156. Make an assembly drawing of the 2'' check valve shown in Fig. 409. Draw a sectional elevation and an exterior end view. This drawing may or may not be dimensioned.

157. The filling-in piece, Fig. 410, is shown one half size. Scale the figure, draw full size, and completely dimension.

158. The guide, Fig. 411, is shown one half size. Scale the figure, draw full size, and completely dimension.

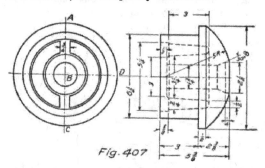

Fig. 407

159. The bracket, Fig. 412, is shown one half size. Scale the figure, draw full size, and completely dimension.

160. The flywheel, Fig. 413, is shown one fourth size. Draw to a scale of 6″ = 1 foot, and completely dimension.

161. The bearing, Fig. 414, is shown one half size. Scale the figure, draw full size, complete the views, and completely dimension.

162. Draw a half end view and a sectional elevation of the pump centerpiece, Fig. 415. Choose a proper scale and completely dimension.

163. Space 5¹/₂″ × 7″. Draw two views of the lever shown in Fig. 416. Both views are to show the true size of the lever.

164. Make detail drawings of each part of the screw stuffing box of Fig. 417. Note that dotted sectioning is used here to indicate the separate pieces. This method is sometimes used to show the exterior and section in the same view.

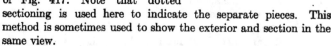

Fig. 408

165. Make an assembly working drawing of the steam jacketed kettle shown in Fig. 418. Draw a half top view and a sectional

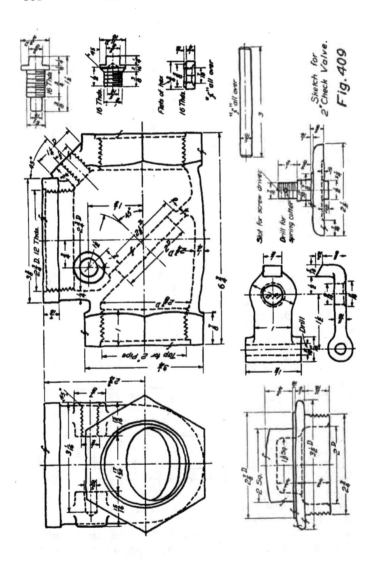

Sketch for
2" Check Valve.

Fig. 409

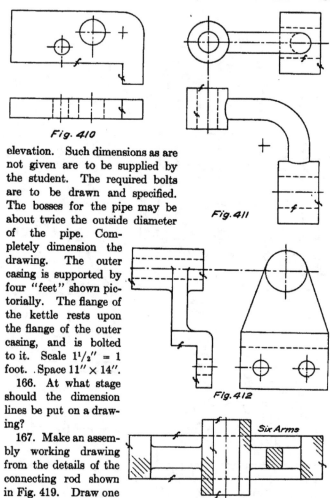

Fig. 410

Fig. 411

Fig. 412

Fig. 413

Six Arms

elevation. Such dimensions as are not given are to be supplied by the student. The required bolts are to be drawn and specified. The bosses for the pipe may be about twice the outside diameter of the pipe. Completely dimension the drawing. The outer casing is supported by four "feet" shown pictorially. The flange of the kettle rests upon the flange of the outer casing, and is bolted to it. Scale $1\frac{1}{2}'' = 1$ foot. Space $11'' \times 14''$.

166. At what stage should the dimension lines be put on a drawing?

167. Make an assembly working drawing from the details of the connecting rod shown in Fig. 419. Draw one view in full and the other half in section and half full. Choose a suitable scale. If necessary a portion of the rod may be broken out. Supply required bolts for wedge

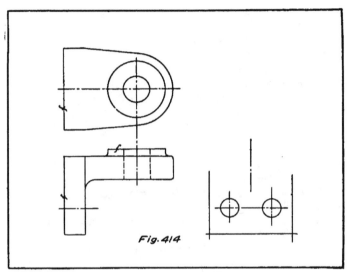

Fig. 414

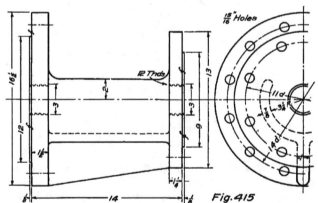

Fig. 415

block. Submit sketch of treatment to instructor for approval. Completely dimension.

168. Make a drawing for the steam cylinder shown in Fig. 420 as follows. Sectional elevation on plane *A–A*; half top view; and section on plane *B–B* looking toward the left. The three

views are to be properly located and completely dimensioned. Show depth of tapped holes. Supply any necessary dimensions that are not given in the figure. Choose a suitable scale.

169. Compute the weight of the Vee block shown in Fig. 335. Tabulate all figures.

170. Compute the weight of the cast iron foot for the steam kettle, Fig. 418. Tabulate all figures.

171. Compute the weight of the outer casing for Fig. 418. (cast iron). Tabulate all figures.

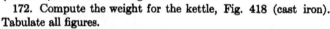

Fig. 416

172. Compute the weight for the kettle, Fig. 418 (cast iron). Tabulate all figures.

173. Compute the weight of the cast iron pulley shown in Fig. 421. Tabulate all figures.

174. How is the diameter of wrought pipe specified?

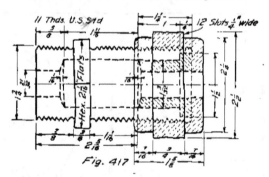

Fig. 417

175. Sketch a 2″ × 2″ × 1¹/₂″ Tee, and mark the size on each opening.

176. Sketch a cross section of a standard pipe thread. Indicate any special features.

177. Draw two views of the piping shown in Fig. 422; one view as shown and the other in the direction of the arrow. Use a double line representation. See Chapter XV.

178. Space 7″ × 11″. Find the curve of intersection between the two cylinders, Fig. 423.

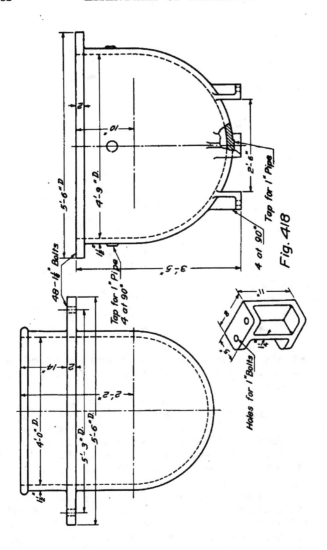

Fig. 418

179. Space 7" × 11". Find the curve of intersection between the two cylinders, Fig. 424.

180. Space 5¹/₂" × 7". Find the intersection between the prisms of Fig. 425.

181. Space 5¹/₂" × 7". Find the intersection between the two prisms of Fig. 426.

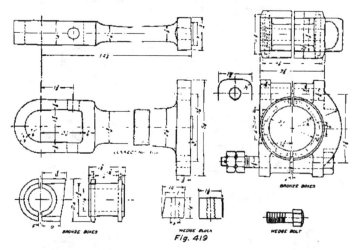

Fig. 419

182. Find the line of intersection between the two cylinders. (Fig. 294, first case.) Both diameters 1¹/₄". Altitude 2¹/₄". Axes intersect.

183. Same as Problem 182 but axes ¹/₂" apart.

184. Find the intersection between two cones (Fig. 294, second case). Diameters 1¹/₈" and altitude 2³/₈". Axes intersect.

185. Same as Problem 184 but with axes ¹/₂" apart.

186. Find the intersection of a cone and a cylinder (Fig. 294 third case). Diameter of cone = 1¹/₂". Diameter of cylinder = 1¹/₄". Axes intersect. Altitude 2¹/₂".

187. Same as Problem 186 but with axes ¹/₂" apart.

188. Find the intersection of a cone and a cylinder (Fig. 294, fourth case). Diameter of cone 3". Altitude of cone 3". Diameter of cylinder 1". Axes intersect.

189. Same as Problem 188 but with axes ¹/₂" apart.

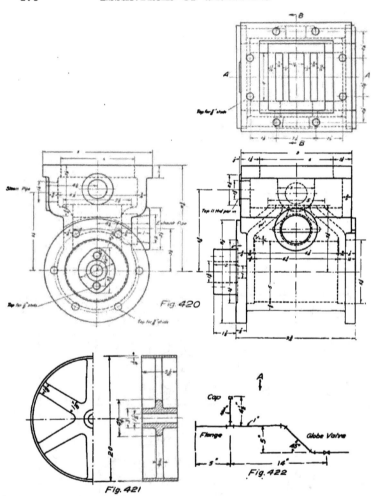

Fig. 420

Fig. 421

Fig. 422

190. Space $5^{1}/_{2}'' \times 7''$. Find the line of intersection between the cone and hexagonal prism of Fig. 427.

191. Space $5^{1}/_{2}'' \times 7''$. Find the line of intersection between the sphere and hexagonal prism of Fig. 428.

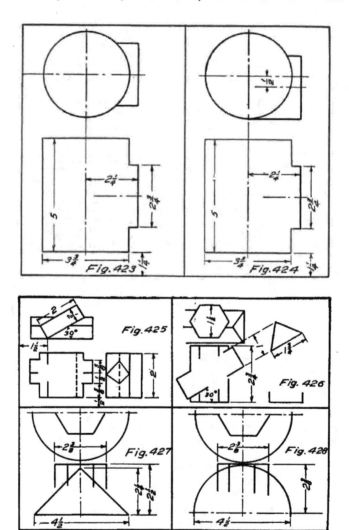

Fig. 423
Fig. 424
Fig. 425
Fig. 426
Fig. 427
Fig. 428

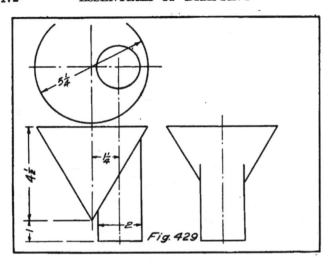

Fig. 429

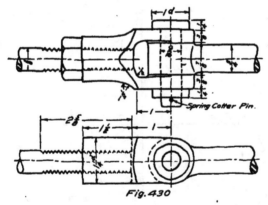

Fig. 430

192. Find the line of intersection between the cone and cylinder of Fig. 429.

193. Make a working drawing of the joint shown in Fig. 430. Find curves accurately.

194. Make a drawing for a connecting rod end (Fig. 295) with dimensions as follows.   Instructor will assign dimensions.

I. $W = 2^1/_2''$ $H = 4''$ $D_1 = 1^1/_2''$
II. $W = 3''$ $H = 3''$ $D_1 = 2''$
III. $W =$ $H =$ $D_1 =$

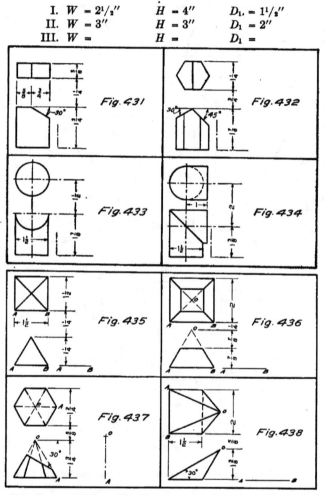

Fig. 431
Fig. 432
Fig. 433
Fig. 434
Fig. 435
Fig. 436
Fig. 437
Fig. 438

195. Space $5^1/_2'' \times 7''$. Develop the lateral surface of the rectangular prism, Fig. 431.

196. Space $5^1/_2'' \times 7''$. Develop the lateral surface of the hexagonal prism, Fig. 432.

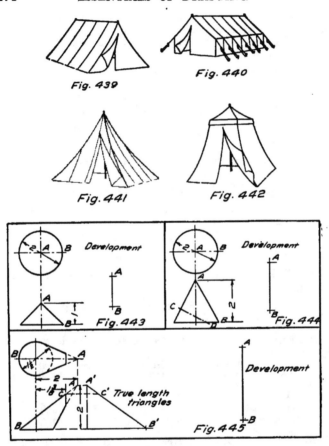

Fig. 439

Fig. 440

Fig. 441

Fig. 442

Fig. 443

Fig. 444

Fig. 445

197. Space $5^1/_2'' \times 7''$. Develop the lateral surface and the upper surface of the cylinder, Fig. 433.

198. Space $5^1/_2'' \times 7''$. Develop the vertical piece of the square elbow, Fig. 434.

199. Space $5^1/_2'' \times 7''$. Develop the lateral surface of the pyramid, Fig. 435.

200. Space $5^1/_2'' \times 7''$. Develop the lateral surface of the frustum of a pyramid, Fig. 436.

201. Space $5^1/_2'' \times 7''$. Develop the lateral surface and the cut face of the hexagonal pyramid, Fig. 437.

202. Space $5^1/_2'' \times 7''$. Develop the lateral surface of the pyramid, Fig. 438.

203. Find the area in square feet of the surface of the tent, Fig. 439.

Size	Width in Feet	Length in Feet	Height in Feet
I	7	7	7
II	9	12	$7^1/_2$

204. Find the area in square feet of the surface of the tent, Fig. 440.

Size	Height of Wall Feet	Length Feet	Width Feet	Total Height Feet
I	3	7	7	7
II	$3^1/_2$	16	12	$7^1/_2$
III	4	20	14	9
IV	5	24	16	11

205. Find the area in square feet of the surface of the tent, Fig. 441.

Size	I	7 feet square	7 feet high
"	II	9 " "	8 " "

206. Find the area in square feet of the surface of the tent, Fig. 442.

Size	Size of Base Feet	Size of Top Feet	Height at Center Feet	Height at Side Feet
I	7 square	$2^1/_2$ square	$7^1/_2$	6
II	8 "	3 "	8	$6^1/_2$
III	10 "	$3^1/_2$ "	9	$7^1/_2$

207. Space $5^1/_2'' \times 7''$. Develop the circular cone shown in Fig. 443. Start with element AB.

208. Space $5^1/_2'' \times 7''$. Develop the part of the surface of cone above the plane CD, Fig. 444. Start with element AB.

209. Develop the portion of a conical surface shown in Fig. 445. First lay out the true length triangles. Then start with element AB.

210. Develop the transition piece of Fig. 446.

211. Develop the transition piece of Fig. 447.

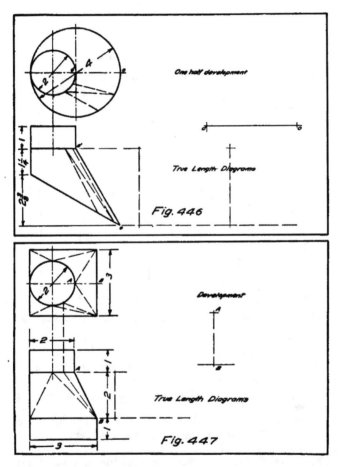

212. Find the intersection of the two cylinders in the three views, Fig. 448. Develop each of the cylinders.

213. Space 5½″ × 7″. Make an isometric drawing of the brass bushing shown in Fig. 175, in section. Full size.

214. Space 11″ × 14″. Make an isometric drawing of the main casting of Fig. 175, in section. Full size.

215. Make an isometric drawing of Fig. 195. Dimensions as furnished by the instructor.

$D_1 = [\ ]\ D_2 = [\ ]\ D_3 = [\ ]\ D_4 = [\ ]\ D_5 = [\ ]\ D_6 = [\ ]$
$L_1 = [\ ]\ L_2 = [\ ]\ L_3 = [\ ]\ L_4 = [\ ]\ L_5 = [\ ]\ L_6 = [\ ]$

216. Space $5^1/_2'' \times 7''$. Make an isometric section of a [   ] diameter rivet and part of two plates each [   ] inches thick.

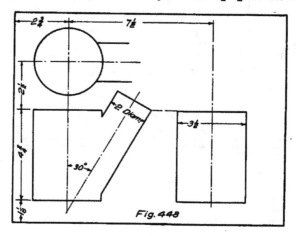

Fig. 448

Dimensions will be furnished by instructor. For forms of rivets see Figs. 133, 134, and 135.

217. Space $5^1/_2'' \times 7''$. Make an isometric drawing of the object shown in Fig. 152. Scale $6'' = 1$ foot.

218. Space $5^1/_2'' \times 7''$. Make a cabinet projection from Fig. 152. Scale $6'' = 1$ foot.

219. Space $5^1/_2'' \times 7''$. For scale of $6'' = 1$ foot. Space $11'' \times 14''$ for full size. Make an isometric drawing of the bearing cap shown in Fig. 174.

220. Space $5^1/_2'' \times 7''$. Make an isometric drawing of Fig. 275. Scale $6'' = 1$ foot.

221. Space $5^1/_2'' \times 7''$. Make an oblique drawing of Fig. 275. Scale $6'' = 1$ foot.

222. Make an oblique drawing of Fig. 273.

223. Make an oblique drawing in section of Fig. 276. Outside diameter $= 4^1/_2''$. Width $= 1$ inch.

224. Space $5^1/_2'' \times 7''$. Make an oblique drawing in section of Fig. 277. Scale $6'' = 1$ foot.

225. Space $5^1/_2'' \times 7''$. Make an isometric drawing in section of Fig. 277.

Shade lines may be used on most any of the problems at the discretion of the instructor.

226. Where should the extra thickness of a shade line be allowed for?

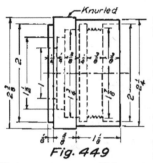

Fig. 449

227. About how wide should the shade lines be compared with the fine lines on a shaded drawing.

228. Make a drawing of Fig. 449, half in section, and half exterior. On the exterior half represent the knurled surface.

229. Refer to Machinery, Power, American Machinist, or other technical papers and make a freehand copy of a simple drawing. Give reference, Paper . . . .

Date . . . . Vol. . . . . . . No. . . . . . . Page . . . . .
Give your criticism, favorable and unfavorable. Consider choice of views; ease of reading and clearness; method of dimensioning; location of dimensions; notes and other information.

Inking Exercises. — Practice exercises are sometimes valuable as a means of teaching accuracy, and for inking practice. The following problems are designed for such purposes. They may be inked with all lines of uniform weight, or with fine and heavy lines as shown. Sharp pencil lines and a minimum of erasing should be insisted upon. When inking, no erasures should be allowed.

230. Exercise 1, Fig. 450. Lay out a $4''$ square. Divide the side AC into 12 equal parts, using the bow spacers. Through each point draw horizontal lines using the T square.

231. Exercise 2, Fig. 450. Lay out a $4''$ square. Divide CD into 12 equal parts with the dividers. Draw vertical lines through each point using a triangle and the T square.

232. Exercise 3, Fig. 450. Lay out a $4''$ square. Divide AC, CD, and BD each into 6 equal parts. Draw BC. Draw lines through the points as shown, using the 45° triangle.

233. Exercise 4, Fig. 450. Lay out a 4″ square. From each corner draw lines making 30° and 60° with the horizontal. Use the 30 × 60 triangle. Stop the lines so as to form the figure shown.

234. Exercise 5, Fig. 450. Lay out a 4″ square. Divide CD and BD into 6 equal parts. Draw lines from point C to each

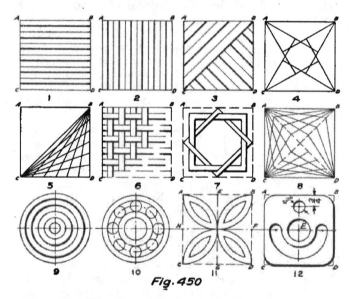

Fig. 450

point on line BD. Draw lines from point B to each point on line CD.

235. Exercise 6, Fig. 450. Lay out a 4″ square. Divide AB and AC each into 12 equal parts. Draw very light horizontal and vertical lines through each point. Brighten up the lines so as to form the figure shown.

236. Exercise 7, Fig. 450. Lay out a 4″ square, a $3^{1}/_{4}″$ square, and a $2^{1}/_{2}″$ square as shown in the figure. Join the middle points of the 4″ square. Join the middle points of the $3^{1}/_{4}″$ square. Erase the lines which are not required.

237. Exercise 8, Fig. 450. Lay out a 4″ square. Draw AD and BC. Divide AD and BC each into 8 equal parts. Join

each point with the corners of the square. When inking be sure to draw toward the corners and allow each line to dry before drawing a second line.

238. Exercise 9, Fig. 450. Draw horizontal and vertical center lines. Using their intersection as a center draw a circle with a diameter of 4″. Divide the radius into 6 equal parts. Through each point thus found draw circles as indicated.

239. Exercise 10, Fig. 450. Draw horizontal and vertical center lines. Draw concentric circles having diameters as follows: 4″, $3^1/_4$″, $2^1/_2$″, $1^3/_4$″, and 1″. Divide the $2^1/_2$″ circle into 8 equal parts and using each point as a center draw small tangent circles having a diameter of $^3/_4$″ as shown in the figure.

240. Exercise 11, Fig. 450. Lay out a 4″ square. Join the middle points of each side by lines HF and EG. Using E, F, G, and H as centers, and a radius of 2″, draw semicircles. Using same centers, and a radius of $1^1/_2$″, draw circle arcs. Erase lines not required to form the figure.

241. Exercise 12, Fig. 450. Lay out a 4″ square. Round the corners with a $^1/_2$″ radius. Find point E, the center of the square. With E as a center, draw a circle having a radius of $^1/_2$″. With E as a center draw two semicircles, having radii of $^3/_4$″ and $1^1/_2$″. Join these semicircles with small circles having a radius of $^3/_8$″. Complete the figure as shown.

INDEX

D. VAN NOSTRAND COMPANY
25 PARK PLACE
NEW YORK

SHORT-TITLE CATALOG

OF

𝕻𝖚𝖇𝖑𝖎𝖈𝖆𝖙𝖎𝖔𝖓𝖘 𝖆𝖓𝖉 𝕴𝖒𝖕𝖔𝖗𝖙𝖆𝖙𝖎𝖔𝖓𝖘

OF

SCIENTIFIC AND ENGINEERING
BOOKS

This list includes
the technical publications of the following English publishers:
SCOTT, GREENWOOD & CO.   JAMES MUNRO & CO., Ltd.
CONSTABLE & COMPANY, Ltd.   TECHNICAL PUBLISHING CO.
ELECTRICIAN PRINTING & PUBLISHING CO.
for whom D. Van Nostrand Company are American agents.

SHORT-TITLE CATALOG

OF THE

Publications and Importations

OF

D. VAN NOSTRAND COMPANY

25 PARK PLACE, N. Y.

All Prices in this list are NET.
All bindings are in cloth unless otherwise noted.

Barnard, J. H. The Naval Militiaman's Guide..........16mo, leather   1 00
Barnard, Major J. G. Rotary Motion. (Science Series No. 90.)....16mo,   0 50
Barnes, J. B. Elements of Military Sketching.................16mo,   *0 60
Barrus, G. H. Engine Tests......................................8vo,   *4 00
Barwise, S. The Purification of Sewage.....................12mo,   3 50
Baterden, J. R. Timber. (Westminster Series.)...............8vo,   *2 00
Bates, E. L., and Charlesworth, F. Practical Mathematics and
    Geometry...12mo,
        Part I. Preliminary and Elementary Course.................   *1 50
        Part II. Advanced Course..................................   *1 50
    —— Practical Mathematics...................................12mo,   *2 00
    —— Practical Geometry and Graphics........................12mo,   2 00
Batey, J. The Science of Works Management.................12mo,   *1 50
    —— Steam Boilers and Combustion...........................12mo,   *1 50
Bayonet Training Manual....................................16mo,   0 30
Beadle, C. Chapters on Papermaking. Five Volumes......12mo, each,   *2 00
Beaumont, R. Color in Woven Design........................8vo,   *6 00
    —— Finishing of Textile Fabrics...........................8vo,   *7 50
    —— Standard Cloths......................................8vo,   *7 50
Beaumont, W. W. The Steam-Engine Indicator...............8vo,   2 50
Bechhold, H. Colloids in Biology and Medicine. Trans. by J. G.
    Bullowa........ ....:(In Press.)
Beckwith, A. Pottery.......................................8vo, paper,   0 60
Bedell, F., and Pierce, C. A. Direct and Alternating Current Manual.
    8vo,   4 00
Beech, F. Dyeing of Cotton Fabrics..........................8vo,   7 50
    —— Dyeing of Woolen Fabrics...............................8vo,   *4 25
Begtrup, J. The Slide Valve.................................8vo,   *2 00
Beggs, G. E. Stresses in Railway Girders and Bridges.....(In Press.)
Bender, C. E. Continuous Bridges. (Science Series No. 26.).....16mo,   0 50
    —— Proportions of Pins used in Bridges. (Science Series No. 4.)
    16mo,   0 50
Bengough, G. D. Brass. (Metallurgy Series.)...........(In Press.)
Bennett, H. G. The Manufacture of Leather...................8vo,   *5 00
Bernthsen, A. A Text-book of Organic Chemistry. Trans. by G.
    M'Gowan 12mo,   *3 00
Bersch, J. Manufacture of Mineral and Lake Pigments. Trans. by A. C.
    Wright...8vo,
Bertin, L. E. Marine Boilers. Trans. by L. S. Robertson.........8vo,   5 00
Beveridge, J. Papermaker's Pocket Book....................12mo,   *4 00
Binnie, Sir A. Rainfall Reservoirs and Water Supply..........8vo,   *3 00
Binns, C. F. Manual of Practical Potting.....................8vo,   *10 00
    —— The Potter's Craft........ :.............................12mo,   *2 00
Birchmore, W. H. Interpretation of Gas Analysis............12mo,   *1 25
Blaine, R. G. The Calculus and Its Applications..............12mo,   *1 75
Blake, W. H. Brewers' Vade Mecum.........................8vo,   *4 00
Blasdale, W. C. Quantitative Chemical Analysis. (Van Nostrand's
    Textbooks.) 12mo,   *2 50
Bligh, W. G. The Practical Design of Irrigation Works 8vo,

Bloch, L.   Science of Illumination.   Trans. by W. C. Clinton......8vo,   *2 50
Blok, A.   Illumination and Artificial Lighting.................12mo,   2 25
Blücher, H.   Modern Industrial Chemistry.   Trans. by J. P. Millington.
                                               8vo,   *7 50
Blyth, A. W.   Foods: Their Composition and Analysis............8vo,   7 50
—— Poisons: Their Effects and Detection.....................8vo,   8 50
Böckmann, F.   Celluloid.....................................12mo,   *3 00
Bodmer, G. R.   Hydraulic Motors and Turbines..............12mo,   5 00
Boileau, J. T.   Traverse Tables............................8vo,   5 00
Bonney, G. E.   The Electro-platers' Handbook.................12mo,   1 50
Booth, N.   Guide to the Ring-spinning Frame................12mo,   *2 00
Booth, W. H.   Water Softening and Treatment.................8vo,   *2 50
—— Superheaters and Superheating and Their Control..........8vo,   *1 50
Bottcher, A.   Cranes: Their Construction, Mechanical Equipment and
      Working.   Trans. by A. Tolhausen....................4to,   *10 00
Bottler, M.   Modern Bleaching Agents.   Trans. by C. Salter....12mo,   *3 00
Bottone, S. R.   Magnetos for Automobilists...................12mo,   *1 00
Boulton, S. B.   Preservation of Timber.   (Science Series No. 82.).16mo,   0 50
Bourcart, E.   Insecticides, Fungicides and Weedkillers..........8vo,   *7 50
Bourgougnon, A.   Physical Problems.   (Science Series No. 113.).16mo,   0 50
Bourry, E.   Treatise on Ceramic Industries.   Trans. by A. B. Searle.
                                               8vo,   *7 25
Bowie, A. J., Jr.   A Practical Treatise on Hydraulic Mining......8vo,   5 00
Bowles, O.   Tables of Common Rocks.   (Science Series No. 125.).16mo,   0 50
Bowser, E. A.   Elementary Treatise on Analytic Geometry......12mo,   1 75
—— Elementary Treatise on the Differential and Integral Calculus.12mo,   2 25
—— Elementary Treatise on Analytic Mechanics..............12mo,   3 00
—— Elementary Treatise on Hydro-mechanics................12mo,   2 50
—— A Treatise on Roofs and Bridges.......................12mo,   *2 25
Boycott, G. W. M.   Compressed Air Work and Diving...........8vo,   *4 25
Bradford, G., 2nd.   Whys and Wherefores of Navigation........12mo,   2 00
—— Sea Terms and Phrases.............12mo, fabrikoid (*In Press.*)
Bragg, E. M.   Marine Engine Design......................12mo,   *2 00
—— Design of Marine Engines and Auxiliaries.................8vo,   *3 00
Brainard, F. R.   The Sextant.   (Science Series No. 101.).......16mo,
Brassey's Naval Annual for 1915.   War Edition.............8vo,   4 00
Briggs, R., and Wolff, A. R.   Steam-Heating.   (Science Series No.
      68.)...16mo,   0 50
Bright, C.   The Life Story of Sir Charles Tilson Bright.........8vo,   *4 50
—— Telegraphy, Aeronautics and War.......................8vo,   6 00
Brislee, T. J.   Introduction to the Study of Fuel.   (Outlines of Indus-
      trial Chemistry.)...................................8vo,   *3 00
Broadfoot, S. K.   Motors: Secondary Batteries.   (Installation Manuals
      Series.)...12mo,   *0 75
Broughton, H. H.   Electric Cranes and Hoists.....................
Brown, G.   Healthy Foundations.   (Science Series No. 80.).....16mo,   0 50
Brown, H.   Irrigation.......................................8vo,   *5 00
Brown, H.   Rubber..8vo,   *2 00
—— W. A.   Portland Cement Industry........................8vo,   3 00
Brown, Wm. N.   Dipping, Burnishing, Lacquering and Bronzing
      Brass Ware 12mo,   *2 00
—— Handbook on Japanning................................12mo,   *2 50

Brown, Wm. N. The Art of Enamelling on Metal............12mo, *2 25
—— House Decorating and Painting...........................12mo, *2 25
—— History of Decorative Art................................12mo, *1 25
—— Workshop Wrinkles 8vo, *1 75
Browne, C. L. Fitting and Erecting of Engines..................8vo, *1 50
Browne, R. E. Water Meters. (Science Series No. 81.).......16mo, 0 50
Bruce, E. M. Detection of Common Food Adulterants........12mo, 1 25
Brunner, R. Manufacture of Lubricants, Shoe Polishes and Leather
          Dressings. Trans. by C. Salter.........................8vo, *4 50
Buel, R. H. Safety Valves. (Science Series No. 21.)..........16mo, 0 50
Bunkley, J. W. Military and Naval Recognition Book.........16mo, 1 00
Burley, G. W. Lathes, Their Construction and Operation......12mo, 2 25
—— Machine and Fitting Shop Practice.....................12mo, 2 50
—— Testing of Machine Tools.............................12mo, 2 50
Burnside, W. Bridge Foundations...........................12mo, *1 50
Burstall, F. W. Energy Diagram for Gas. With Text...........8vo, 1 50
—— Diagram. Sold separately *1 00
Burt, W. A. Key to the Solar Compass.............16mo, leather, 2 50
Buskett, E. W. Fire Assaying..............................12mo, *1 25
Butler, H. J. Motor Bodies and Chassis......................8vo, *3 00
Byers, H. G., and Knight, H. G. Notes on Qualitative Analysis....8vo, *1 50
Cain, W. Brief Course in the Calculus.......................12mo, *1 75
—— Elastic Arches. (Science Series No. 48.).................16mo, 0 50
—— Maximum Stresses. (Science Series No. 38.)............16mo, 0 50
—— Practical Designing Retaining of Walls. (Science Series No. 3.)
                                                              16mo, 0 50
—— Theory of Steel-concrete Arches and of Vaulted Structures.
          (Science Series No. 42.)............................16mo, 0 50
—— Theory of Voussoir Arches. (Science Series No. 12.).....16mo, 0 50
—— Symbolic Algebra. (Science Series No. 73.)..............16mo, 0 50
Calvert, G. T. The Manufacture of Sulphate of Ammonia and
          Crude Ammonia 12mo, 4 00
Carpenter, F. D. Geographical Surveying. (Science Series No. 37.).16mo,
Carpenter, R. C., and Diederichs, H. Internal Combustion Engines..8vo, *5 00
Carter, H. A. Ramie (Rhea), China Grass....................12mo, *3 00
Carter, H. R. Modern Flax, Hemp, and Jute Spinning..........8vo, *4 50
—— Bleaching, Dyeing and Finishing of Fabrics..............8vo, *1 25
Cary, E. R. Solution of Railroad Problems with the Slide Rule..16mo, *1 00
Casler, M. D. Simplified Reinforced Concrete Mathematics......12mo, *1 00
Cathcart, W. L. Machine Design. Part I. Fastenings..........8vo, *3 00
Cathcart, W. L., and Chaffee, J. I. Elements of Graphic Statics..,8vo, *3 00
—— Short Course in Graphics...............................12mo, 1 50
Caven, R. M., and Lander, G. D. Systematic Inorganic Chemistry.12mo, *2 00
Chalkley, A. P. Diesel Engines..............................8vo, *4 00
Chalmers. T. W. The Production and Treatment of Vegetable Oils,
                                                              4to, 7 50
Chambers' Mathematical Tables.............................8vo, 1 75
Chambers, G. F. Astronomy................................16mo, *1 50
Chappel, E. Five Figure Mathematical Tables..................8vo, *2 00
Charnock, Mechanical Technology..........................8vo, *3 00
Charpentier, P. Timber....................................8vo, *7 25
Chatley, H. Principles and Designs of Aeroplanes. (Science Series
          No. 126)...16mo, 0 50
—— How to Use Water Power...............................12mo, *1 50
—— Gyrostatic Balancing 8vo, *1 25

Garcia, A. J. R. V. Spanish-English Railway Terms.............8vo, *4 50
Gardner, H. A. Paint Researches, and Their Practical Applications,
                                                         8vo, *5 00
Garforth, W. E. Rules for Recovering Coal Mines after Explosions and
        Fires....................................12mo, leather,   1 50
Garrard, C. C. Electric Switch and Controlling Gear.............8vo, *6 00
Gaudard, J. Foundations. (Science Series No. 34.)...........16mo,   0 50
Gear, H. B., and Williams, P. F. Electric Central Station Distribution
        Systems 8vo, *3 50
Geerligs, H. C. P. Cane Sugar and Its Manufacture.............8vo, *6 00
—— Chemical Control in Cane Sugar Factories.................4to,   5 00
Geikie, J. Structural and Field Geology.................... 8vo, *4 00
—— Mountains. Their Growth, Origin and Decay.............8vo, *4 00
—— The Antiquity of Man in Europe.......................8vo, *3 00
Georgi, F., and Schubert, A. Sheet Metal Working. Trans. by C.
        Salter............................... ....8vo,   4 25
Gerhard, W. P. Sanitation, Watersupply and Sewage Disposal of Country
        Houses.......................................12mo, *2 00
—— Gas Lighting (Science Series No. 111.)..................16mo,   0 50
—— Household Wastes. (Science Series No. 97.)..............16mo,   0 50
—— House Drainage. (Science Series No. 63.)................16mo,   0 50
—— Sanitary Drainage of Buildings. (Science Series No. 93.) 16mo,   0 50
Gerhardi, C. W. H. Electricity Meters.........................8vo, *6 00
Geschwind, L. Manufacture of Alum and Sulphates. Trans. by C.
        Salter...................................... ....8vo, *5 00
Gibbings, A. H. Oil Fuel Equipment for Locomotives..........8vo, *2 50
Gibbs, W. E. Lighting by Acetylene..........................12mo, *1 50
Gibson, A. H. Hydraulics and Its Application.................8vo, *5 00
—— Water Hammer in Hydraulic Pipe Lines.................12mo, *2 00
Gibson, A. H., and Ritchie, E. G. Circular Arc Bow Girder........4to, *3 50
Gilbreth, F. B. Motion Study................................12mo, *2 00
—— Bricklaying System 8vo, *3 00
—— Field System ....12mo, leather, *3 00
—— Primer of Scientific Management................... ..12mo, *1 00
Gillette, H. P. Handbook of Cost Data.................12mo, leather, *5 00
—— Rock Excavation Methods and Cost.....................12mo, *5 00
—— and Dana, R. T. Cost Keeping and Management Engineering.8vo, *3 50
—— and Hill, C. S. Concrete Construction, Methods and Cost....8vo, *5 00
Gillmore, Gen. Q. A. Roads, Streets, and Pavements............12mo,   1 25
Godfrey, E. Tables for Structural Engineers..........16mo, leather, *2 50
Golding, H. A. The Theta-Phi Diagram......................12mo, *2 00
Goldschmidt, R. Alternating Current Commutator Motor..........8vo, *3 00
Goodchild, W. Precious Stones. (Westminster Series.)....... ...8vo, *2 00
Goodell, J. M. The Location, Construction and Maintenance of
        Roads 8vo,   1 00
Goodeve, T. M. Textbook on the Steam-engine.................12mo,   2 00
Gore, G. Electrolytic Separation of Metals......... ....8vo, *3 50
Gould, E. S. Arithmetic of the Steam-engine..................12mo,   1 00
—— Calculus. (Science Series No. 112.)....................16mo,   0 50
—— High Masonry Dams. (Science Series No. 22.)............16mo,   0 50
Gould, E. S. Practical Hydrostatics and Hydrostatic Formulas. (Science
        Series No. 117.)...............................16mo,   0 50

Gratacap, L. P. A Popular Guide to Minerals..................8vo,   *2 00
Gray, J. Electrical Influence Machines 12mo,   2 00
——— Marine Boiler Design.....................................12mo,   *1 25
Greenhill, G. Dynamics of Mechanical Flight..................8vo,   *2 50
Gregorius, R. Mineral Waxes. Trans. by C. Salter..........12mo,   *3 50
Grierson, R. Some Modern Methods of Ventilation..............8vo,   *3 00
Griffiths, A. B. A Treatise on Manures......................12mo,   3 00
——— Dental Metallurgy 8vo,   *4 25
Gross, E. Hops..8vo,   *6 25
Grossman, J. Ammonia and Its Compounds................... ..12mo,   *1 25
Groth, L. A. Welding and Cutting Metals by Gases or Electricity.
      (Westminster Series)......................:............8vo,   *2 00
Grover, F. Modern Gas and Oil Engines.......................8vo,   *3 00
Gruner, A. Power-loom Weaving...............................8vo,   *3 00
Grunsky, C. E. Topographic Stadia Surveying................16mo,   2 00
Güldner, Hugo. Internal Combustion Engines. Trans. by H. Diederichs.
                                                            4to,   *15 00
Gunther, C. O. Integration..................................8vo,   *1 25
Gurden, R. L. Traverse Tables..................folio, half morocco,   *7 50
Guy, A. E. Experiments on the Flexure of Beams...............8vo,   *1 25

Haenig, A. Emery and Emery Industry.........................8vo,   *3 00
Hainbach, R. Pottery Decoration. Trans., by C. Salter........12mo,   *4 25
Hale, W. J. Calculations of General Chemistry...............12mo,   *1 25
Hall, C. H. Chemistry of Paints and Paint Vehicles..........12mo,   *2 00
Hall, G. L. Elementary Theory of Alternate Current Working....8vo,
Hall, R. H. Governors and Governing Mechanism..............12mo,   *2 50
Hall, W. S. Elements of the Differential and Integral Calculus......8vo,   *2 25
——— Descriptive Geometry ....8vo volume and a 4to atlas,   *3 50
Haller, G. F., and Cunningham, E. T. The Tesla Coil...........12mo,   *1 25
Halsey, F. A. Slide Valve Gears.............................12mo,   1 50
——— The Use of the Slide Rule. (Science Series No. 114.).........16mo,   0 50
——— Worm and Spiral Gearing. (Science Series No. 116.)........16mo,   0 50
Hancock, H. Textbook of Mechanics and Hydrostatics......... ..8vo,   1 50
Hancock, W. C. Refractory Materials. (Metallurgy Series.) (In Press.)
Hardy, E. Elementary Principles of Graphic Statics............12mo,   *1 50
Haring, H. Engineering Law.
      Vol. I. Law of Contract...................................8vo,   *4 00
Harper, J. H. Hydraulic Tables on the Flow of Water........16mo,   *2 00
Harris, S. M. Practical Topographical Surveying.........(In Press.)
Harrison, W. B. The Mechanics' Tool-book....................12mo,   1 50
Hart, J. W. External Plumbing Work..........................8vo,   *3 25
——— Hints to Plumbers on Joint Wiping......................8vo,   *4 25
——— Principles of Hot Water Supply.........................8vo,   *4 25
——— Sanitary Plumbing and Drainage........................8vo,   *4 25
Haskins, C. H. The Galvanometer and Its Uses................16mo,   1 50
Hatt, J. A. H. The Colorist.........................square 12mo,   *1 50
Hausbrand, E. Drying by Means of Air and Steam. Trans. by A. C.
      Wright 12mo,   *3 00
——— Evaporating, Condensing and Cooling Apparatus. Trans. by A. C.
      Wright...... ....8vo,   *7 25

Hausmann, E. Telegraph Engineering...........................8vo,   *3 00
Hausner, A. Manufacture of Preserved Foods and Sweetmeats. Trans.
    by A. Morris and H. Robson.............................8vo,   *4 25
Hawkesworth, J. Graphical Handbook for Reinforced Concrete Design.
                                            4to,   *2 50
Hay, A. Continuous Current Engineering......................8vo,   *2 50
Hayes, H. V. Public Utilities, Their Cost New and Depreciation...8vo,   *2 00
——— Public Utilities, Their Fair Present Value and Return......8vo,   *2 00
Heath, F. H. Chemistry of Photography..............8vo. (*In Press.*)
Heather, H. J. S. Electrical Engineering.....................8vo,   *3 50
Heaviside, O. Electromagnetic Theory. Vols. I and II....8vo, each,   *5 00
                              Vol. III..............8vo,   *7 50
Heck, R. C. H. The Steam Engine and Turbine................8vo,   *3. 50
——— Steam-Engine and Other Steam Motors. Two Volumes.
    Vol. I. Thermodynamics and the Mechanics................8vo,   *3 50
    Vol. II. Form, Construction, and Working...................8vo,   *5 00
——— Notes on Elementary Kinematics...................8vo, boards,   *1 00
— — Graphics of Machine Forces.....................8vo, boards,   *1 00
Heermann, P. Dyers' Materials. Trans. by A. C. Wright........12mo,   *3 00
Heidenreich, E. L. Engineers' Pocketbook of Reinforced Concrete,
                                16mo, leather,   *3 00
Hellot, Macquer and D'Apligny. Art of Dyeing Wool, Silk and Cotton. 8vo,   *2 00
Henrici, O. Skeleton Structures...............................8vo,   1 50
Hering, C., and Getman, F. H. Standard Tables of Electro-Chemical
    Equivalents 12mo,   *2 00
Hering, D. W. Essentials of Physics for College Students........8vo,   *1 75
Hering-Shaw, A. Domestic Sanitation and Plumbing. Two Vols...8vo,   *5 00
Hering-Shaw, A. Elementary Science 8vo,   *2 00
Herington, C. F. Powdered Coal as Fuel......................8vo,   3 00
Herrmann, G. The Graphical Statics of Mechanism. Trans. by A. P.
    Smith...12mo,   2 00
Herzfeld, J. Testing of Yarns and Textile Fabrics..............8vo,   *6 25
Hildebrandt, A. Airships, Past and Present.....................8vo,
Hildenbrand, B. W. Cable-Making. (Science Series No. 32.)....16mo,   0 50
Hilditch, T. P. A Concise History of Chemistry..............12mo,   *1 25
Hill, C. S. Concrete Inspection............................16mo,   *1 00
Hill, J. W. The Purification of Public Water Supplies. New Edition.
                                    (*In Press.*)
——— Interpretation of Water Analysis...................(*In Press.*)
Hill, M. J. M. The Theory of Proportion.......................8vo,   *2 50
Hiroi, I. Plate Girder Construction. (Science Series No. 95.)...16mo,   0 50
——— Statically-Indeterminate Stresses........ ....12mo,   *2 00
Hirshfeld, C. F. Engineering Thermodynamics. (Science Series No. 45.)
                                    16mo,   0 50
Hoar, A. The Submarine Torpedo Boat.......................12mo,   *2 00
Hobart, H. M. Heavy Electrical Engineering....................8vo,   *4 50
——— Design of Static Transformers..........................12mo,   *2 00
——— Electricity........ ....8vo,   *2 00
——— Electric Trains........ ....8vo,   *2 50
——— Electric Propulsion of Ships.............................8vo,   *2 50

Inness, C. H. Problems in Machine Design....................12mo,   *3 00
—— Air Compressors and Blowing Engines....................12mo,
—— Centrifugal Pumps 12mo,   *3 00
—— The Fan 12mo,   *4 00

Jacob, A., and Gould, E. S. On the Designing and Construction of
    Storage Reservoirs. (Science Series No. 6)...........16mo,   0 50
Jannettaz, E. Guide to the Determination of Rocks. Trans. by G. W.
    Plympton....................................,...........12mo,   1 50
Jehl, F. Manufacture of Carbons.............................8vo,   *4 00
Jennings, A. S. Commercial Paints and Painting. (Westminster Series.)
                                     8vo,   *4 00
Jennison, F. H. The Manufacture of Lake Pigments............8vo,   *3 00
Jepson, G. Cams and the Principles of their Construction........8vo,   *1 50
—— Mechanical Drawing....................8vo (In Preparation.)
Jervis-Smith, F. J. Dynamometers...........................8vo,   *3 50
Jockin, W. Arithmetic of the Gold and Silversmith...........12mo,   *1 00
Johnson, J. H. Arc Lamps and Accessory Apparatus. (Installation
    Manuals Series.)....................................12mo,   *0 75
Johnson, T. M. Ship Wiring and Fitting. (Installation Manuals Series.)
                   12mo,   *0 75
Johnson, W. McA. The Metallurgy of Nickel...... (In Preparation.)
Johnston, J. F. W., and Cameron, C. Elements of Agricultural Chemistry
    and Geology....................................12mo,   2 60
Joly, J. Radioactivity and Geology..........................12mo,   *3 00
Jones, H. C. Electrical Nature of Matter and Radioactivity 12mo,   *2 00
—— Nature of Solution.......................................8vo,   *3 50
—— New Era in Chemistry....................................12mo,   *2 00
Jones, J. H. Tinplate Industry.............................8vo,   *3 00
Jones, M. W. Testing Raw Materials Used in Paint..........12mo,   *3 00
Jordan, L. C. Practical Railway Spiral.............12mo, leather,   *1 50
Joynson, F. H. Designing and Construction of Machine Gearing ..8vo,   2 00
Jüptner, H. F. V. Siderology: The Science of Iron.............8vo,   *6 25

Kapp, G. Alternate Current Machinery. (Science Series No. 96.).16mo,   0 50
Kapper, F. Overhead Transmission Lines...................4to,   *4 00
Keim, A. W. Prevention of Dampness in Buildings.............8vo,   *3 00
Keller, S. S. Mathematics for Engineering Students. 12mo, half leather.
—— and Knox, W. E. Analytical Geometry and Calculus............   *2 00
Kelsey, W. R. Continuous-current Dynamos and Motors.........8vo,   *2 50
Kemble, W. T., and Underhill, C. R. The Periodic Law and the Hydrogen
    Spectrum....................................8vo, paper,   *0 50
Kemp, J. F. Handbook of Rocks............................8vo,   *1 50
Kennedy, A. B. W., and Thurston, R. H. Kinematics of Machinery.
    (Science Series No. 54.)..........................16mo,   0 50
Kennedy, A. B. W., Unwin, W. C., and Idell, F. E. Compressed Air.
    (Science Series No. 106.)..........................16mo,   0 50

Kennedy, R. Electrical Installations. Five Volumes............4to,   15 00
    Single Volumes ....each,   3 50
—— Flying Machines; Practice and Design...................12mo,   *2 50
—— Principles of Aeroplane Construction.......................8vo,   *2 00
Kennelly, A. E. Electro-dynamic Machinery....................8vo,   1 50
Kent, W. Strength of Materials. (Science Series No. 41.).....16mo,   0 50
Kershaw, J. B. C. Fuel, Water and Gas Analysis...............8vo,   *2 50
—— Electrometallurgy. (Westminster Series.).................8vo,   *2 00
—— The Electric Furnace in Iron and Steel Production........12mo,
—— Electro-Thermal Methods of Iron and Steel Production....8vo,   *3 00
Kindelan, J. Trackman's Helper..............................12mo,   2 00
Kinzbrunner, C. Alternate Current Windings.................8vo,   *1 50
—— Continuous Current Armatures............................8vo,   *1 50
—— Testing of Alternating Current Machines 8vo,   *2 00
Kirkaldy, A. W., and Evans, A. D. History and Economics of
    Transport 8vo,   *3 00
Kirkaldy, W. G. David Kirkaldy's System of Mechanical Testing..4to,   10 00
Kirkbride, J. Engraving for Illustration.......................8vo,   *1 75
Kirkham, J. E. Structural Engineering........................8vo,   *5 00
Kirkwood, J. P. Filtration of River Waters.....................4to,   7 50
Kirschke, A. Gas and Oil Engines............................12mo,   *1 50
Klein, J. F. Design of a High-speed Steam-engine 8vo,   *5 00
—— Physical Significance of Entropy...........................8vo,   *1 50
Klingenberg, G. Large Electric Power Stations.................4to,   *5 00
Knight, R.-Adm. A. M. Modern Seamanship..................8vo,   *6 50
—— Pocket Edition...............................12mo, fabrikoid,   3 00
Knott, C. G., and Mackay, J. S. Practical Mathematics.........8vo,   2 00
Knox, G. D. Spirit of the Soil...............................12mo,   *1 25
Knox, J. Physico-Chemical Calculations.......................12mo,   *1 25
—— Fixation of Atmospheric Nitrogen. (Chemical Monographs.).12mo,   *1 00
Koester, F. Steam-Electric Power Plants......................4to,   *5 00
—— Hydroelectric Developments and Engineering................4to,   *5 00
Koller, T. The Utilization of Waste Products..................8vo,   *6 50
—— Cosmetics...8vo,   *3 00
Koppe, S. W. Glycerine.....................................12mo,   *4 25
Kozmin, P. A. Flour Milling. Trans. by M. Falkner..........8vo,   7 50
Kremann, R. Application of the Physico-Chemical Theory to Tech-
    nical Processes and Manufacturing Methods. Trans. by H.
    E. Potts..8vo,   *3 00
Kretchmar, K. Yarn and Warp Sizing.....................8vo,   *6 25

Laffargue, A. Attack in Trench Warfare.....................16mo,   0 50
Lallier, E. V. Elementary Manual of the Steam Engine........12mo,   *2 00
Lambert, T. Lead and Its Compounds........................8vo,   *4 25
—— Bone Products and Manures.............................8vo,   *4 25
Lamborn, L. L. Cottonseed Products.........................8vo,   *3 00
—— Modern Soaps, Candles, and Glycerin.....................8vo,   *7 50
Lamprecht, R. Recovery Work After Pit Fires. Trans. by C. Salter.8vo,   *6 25
Lancaster, M. Electric Cooking, Heating and Cleaning..........8vo,   *1 00
Lanchester, F. W. Aerial Flight. Two Volumes. 8vo.
    Vol. I. Aerodynamics...   *6 00
    Vol. II. Aerodonetics..   *6 00

Lodge, O. J. Elementary Mechanics. 12mo,   1 50
—— Signalling Across Space without Wires................8vo,   *2 00
Loewenstein, L. C., and Crissey, C. P. Centrifugal Pumps...........   *4 50
Lomax, J. W. Cotton Spinning..............................12mo,   1 50
Lord, R. T. Decorative and Fancy Fabrics....................8vo,   *4 25
Loring, A. E. A Handbook of the Electromagnetic Telegraph....16mo   0 50
——— Handbook. (Science Series No. 39.)....................16m,   0 50°
Lovell, D. H. Practical Switchwork..........................12mo,   *1 00
Low, D. A. Applied Mechanics (Elementary)................16mo,   0 80
Lubschez, B. J. Perspective................................12mo,   *1 50
Lucke, C. E. Gas Engine Design............................8vo,   *3 00
—— Power Plants: Design, Efficiency, and Power Costs. 2 vols.
(In Preparation.)
Luckiesh, M. Color and Its Application......................8vo,   *3 00
——— Light and Shade and Their Applications.................8vo,   *2 50
Lunge, G. Coal-tar and Ammonia. Three Volumes.............8vo,   *25 00
—— Technical Gas Analysis...................................8vo,   *4 50
—— Manufacture of Sulphuric Acid and Alkali. Four Volumes....8vo,
Vol. I. Sulphuric Acid. In three parts.....................   *18 00
——Vol. I. Supplement.......................................8vo,   5 00
Vol. II. Salt Cake, Hydrochloric Acid and Leblanc Soda. In two
parts ....(In Press.)
Vol. III. Ammonia Soda............................(In Press.)
Vol. IV. Electrolytic Methods...................:.......(In Press.)
—— Technical Chemists' Handbook...................12mo, leather,   *4 00
—— Technical Methods of Chemical Analysis. Trans. by C. A. Keane
in collaboration with the corps of specialists.
Vol.   I. In two parts.:................................8vo,   *15 00
Vol. II. In two parts................................8vo,   *18 00
Vol. III. In two parts...............................8vo,   *18 00
The set (3 vols.) complete..................................   *50 00
Luquer, L. M. Minerals in Rock Sections...................8vo,   *1 50

MacBride, J. D. A Handbook of Practical Shipbuilding,
12mo, fabrikoid (In Press.)
Macewen, H. A. Food Inspection...........................8vo,   *2 50
Mackenzie, N. F. Notes on Irrigation Works..................8vo,   *2 50
Mackie, J. How to Make a Woolen Mill Pay..................8vo,   *2 25
Maguire, Wm. R. Domestic Sanitary Drainage and Plumbing 8vo,   4 00
Malcolm, C. W. Textbook on Graphic Statics...................8vo,   *3 00
Malcolm, H. W. Submarine Telegraph Cable..............(In Press.)
Mallet, A. Compound Engines. Trans. by R. R. Buel. (Science Series
No. 10.)..16mo,
Mansfield, A. N. Electro-magnets. (Science Series No. 64.) 16mo,   0 50
Marks, E. C. R. Construction of Cranes and Lifting Machinery..12mo,   *2 00
—— Construction and Working of Pumps...................12mo,
—— Manufacture of Iron and Steel Tubes...................12mo,   *2 00
——Mechanical Engineering Materials......................12mo,   *1 50
Marks, G. C. Hydraulic Power Engineering...................8vo,   4 50
—— Inventions, Patents and Designs.......................12mo,   *1 00
Marlow, T. G. Drying Machinery and Practice................8vo,   *5 00

Marsh, C. F. Concise Treatise on Reinforced Concrete 8vo,   *2 50
—— Reinforced Concrete Compression Member Diagram. Mounted on
    Cloth Boards..   *1.50
Marsh, C. F., and Dunn, W. Manual of Reinforced Concrete and Con-
    crete Block Construction........16mo, fabrikoid (*In Press*.)
Marshall, W. J., and Sankey, H. R. Gas Engines. (Westminster Series.)
        8vo,   *2 00
Martin, G. Triumphs and Wonders of Modern Chemistry.......8vo,   *2 00
—— Modern Chemistry and Its Wonders........................8vo,   *2 00
Martin, N. Properties and Design of Reinforced Concrete......12mo,   *2 50
Martin, W. D. Hints to Engineers...........................12mo,   *1 50
Massie, W. W., and Underhill, C. R. Wireless Telegraphy and Telephony.
        12mo,   *1 00
Mathot, R. E. Internal Combustion Engines....................8vo,   *4 00
Maurice, W. Electric Blasting Apparatus and Explosives........8vo,   *3 50
—— Shot Firer's Guide.......................................8vo,   *1 50
Maxwell, F. Sulphitation in White Sugar Manufacture.......12mo,   3 75
Maxwell, J. C. Matter and Motion. (Science Series No. 36.).
        16mo,   0 50
Maxwell, W. H., and Brown, J. T. Encyclopedia of Municipal and Sani-
    tary Engineering...4to,   *10 00
Mayer, A. M. Lecture Notes on Physics......................8vo,   2 00
Mayer, C., and Slippy, J. C. Telephone Line Construction........8vo,   *3 00
McCullough, E. Practical Surveying.........................12mo,   *2 00
—— Engineering Work in Cities and Towns....................8vo,   *3 00
—— Reinforced Concrete 12mo,   *1 50
McCullough, R. S. Mechanical Theory of Heat...............8vo,   3 50
McGibbon, W. C. Indicator Diagrams for Marine Engineers.......8vo,   *3 50
—— Marine Engineers' Drawing Book....................oblong 4to,   *2 50
McGibbon, W. C. Marine Engineers Pocketbook...............12mo,   *4 00
McIntosh, J. G. Technology of Sugar........................8vo,   *7 25
—— Industrial Alcohol 8vo,   *4 25
—— Manufacture of Varnishes and Kindred Industries. Three Volumes.
    8vo.
      Vol. I. Oil Crushing, Refining and Boiling....................
      Vol. II. Varnish Materials and Oil Varnish Making... ....   *6 25
      Vol. III. Spirit Varnishes and Materials.....................   *7 25
McKay, C. W. Fundamental Principles of the Telephone Business.
        8vo. (*In Press*.)
McKillop, M., and McKillop, A. D. Efficiency Methods.........12mo,   1 50
McKnight, J. D., and Brown, A. W. Marine Multitubular Boilers....   *2 50
McMaster, J. B. Bridge and Tunnel Centres. (Science Series No. 20.)
        16mo,   0 50
McMechen, F. L. Tests for Ores, Minerals and Metals........12mo,   *1 00
McPherson, J. A. Water-works Distribution...................8vo,   2 5c
Meade, A. Modern Gas Works Practice.......................8vo,   *8 50
Meade, R. K. Design and Equipment of Small Chemical Laboratories,
        8vo,
Melick, C. W. Dairy Laboratory Guide.......................12mo,   *1 25
Mensch, L. J. Reinforced Concrete Pocket Book........16mo, leather,   *4 00
Merck, E. Chemical Reagents; Their Purity and Tests. Trans. by
    H. E. Schenck...8vo,   1 00
Merivale, J. H. Notes and Formulae for Mining Students.....12mo,   1 50
Merritt, Wm. H. Field Testing for Gold and Silver....16mo, leather,   2 00

Mertens. Tactics and Technique of River Crossings. Translated by
    W. Kruger...8vo,   2 50
Mierzinski, S. Waterproofing of Fabrics. Trans. by A. Morris and H.
    Robson...8vo,   *3 00
Miessner, B. F. Radio Dynamics.........................12mo,   *2 00
Miller, G. A. Determinants. (Science Series No 105.)........16mo,
Miller, W. J. Introduction to Historical Geology.............12mo,   *2 00
Milroy, M. E. W. Home Lace-making........................12mo,   *1 00
Mills, C. N. Elementary Mechanics for Engineers.............8vo,   *1 00
Mitchell, C. A. Mineral and Aerated Waters................8vo,   *3 00
Mitchell, C. A., and Prideaux, R. M. Fibres Used in Textile and Allied
    Industries...8vo,   *3 00
Mitchell, C. F., and G. A. Building Construction and Drawing. 12mo.
    Elementary Course..   *1 50
    Advanced Course...   *2 50
Monckton, C. C. F. Radiotelegraphy. (Westminster Series.).....8vo,   *2 00
Monteverde, R. D. Vest Pocket Glossary of English-Spanish, Spanish-
    English Technical Terms....................64mo, leather,   *1 00
Montgomery, J. H. Electric Wiring Specifications.............16mo,   *1 00
Moore, E. C. S. New Tables for the Complete Solution of Ganguillet and
    Kutter's Formula...8vo,   *5 00
Moore, Harold. Liquid Fuel for Internal Combustion Engines...8vo,   5 00
Morecroft, J. H., and Hehre, F. W. Short Course in Electrical Testing.
                                                 8vo,   *1 50
Morgan, A. P. Wireless Telegraph Apparatus for Amateurs......12mo,   *1 50
Morgan, C. E. Practical Seamanship for the Merchant Marine,
                     12mo, fabrikoid (*In Press.*)
Moses, A. J. The Characters of Crystals......................8vo,   *2 00
—— and Parsons, C. L. Elements of Mineralogy................8vo,   *3 00
Moss, S. A. Elements of Gas Engine Design.(Science Series No.121.)16mo,   0 50
—— The Lay-out of Corliss Valve Gears. (Science Series No. 119.)16mo,   0 50
Mulford, A. C. Boundaries and Landmarks....................12mo,   *1 00
Mullin, J. P. Modern Moulding and Pattern-making..........12mo,   2 50
Munby, A. E. Chemistry and Physics of Building Materials. (West-
    minster Series.)...8vo,   *2 00
Murphy, J. G. Practical Mining............................16mo,   1 00
Murray, J. A. Soils and Manures. (Westminster Series.).......8vo,   *2 00

Nasmith, J. The Student's Cotton Spinning..................8vo,   3 00
—— Recent Cotton Mill Construction........................12mo,   2 50
Neave, G. B., and Heilbron, I. M. Identification of Organic Compounds.
                     12mo,   °1 25
Neilson, R. M. Aeroplane Patents..........................8vo,   *2 00
Nerz, F. Searchlights. Trans. by C. Rodgers................8vo,   *3 00
Neuberger, H., and Noalhat, H. Technology of Petroleum. Trans. by
    J. G. McIntosh...8vo,   *10 00
Newall, J. W. Drawing, Sizing and Cutting Bevel-gears.........8vo,   1 50
Newbigin, M. I., and Flett, J. S. James Geikie, the Man and the
    Geologist..8vo,   3 50
Newbeging, T. Handbook for Gas Engineers and Managers.....8vo,   *6 50
Newell, F. H., and Drayer, C. E. Engineering as a Career..12mo, cloth,   *1 00
                                                  paper,   0 75
Nicol, G. Ship Construction and Calculations................8vo,   *5 00
Nipher, F. E. Theory of Magnetic Measurements.............12mo,   1 00

Nisbet, H. Grammar of Textile Design.........................8vo,
Nolan, H. The Telescope. (Science Series No. 51.)...........16mo,   0 50
Norie, J. W. Epitome of Navigation (2 Vols.).................octavo, 15 00
—— A Complete Set of Nautical Tables with Explanations of Their
        Use ....octavo,   6 50
North, H. B. Laboratory Experiments in General Chemistry.....12mo,   *1 00
Nugent, E. Treatise on Optics...............................12mo,   1 50

O'Connor, H. The Gas Engineer's Pocketbook.........12mo, leather,   3 50
Ohm, G. S., and Lockwood, T. D. Galvanic Circuit. Translated by
        William Francis. (Science Series No. 102.)...........16mo,   0 50
Olsen, J. C. Text-book of Quantitative Chemical Analysis......8vo,   3 50
Olsson, A. Motor Control, in Turret Turning and Gun Elevating. (U. S.
        Navy Electrical Series, No. 1.).................12mo, paper, *0 50
Ormsby, M. T. M. Surveying..................................12mo   2 50
Oudin, M. A. Standard Polyphase Apparatus and Systems......8vo,   *3 00
Owen, D. Recent Physical Research...........................8vo,

Pakes, W. C. C., and Nankivell, A. T. The Science of Hygiene ..8vo, *1 75
Palaz, A. Industrial Photometry. Trans. by G. W. Patterson, Jr..8vo, *4 00
Pamely, C. Colliery Manager's Handbook......................8vo, *10 00
Parker, P. A. M. The Control of Water........................8vo,   *5 00
Parr, G. D. A. Electrical Engineering Measuring Instruments....8vo, *3 50
Parry, E. J. Chemistry of Essential Oils and Artificial Perfumes.... 10 00
—— Foods and Drugs. Two Volumes.
        Vol. I. Chemical and Microscopical Analysis of Foods and Drugs. *10 00
        Vol. II. Sale of Food and Drugs Act.......................... *4 25
—— and Coste, J. H. Chemistry of Pigments.....................8vo,   *6 50
Parry, L. Notes on Alloys..8vo,   *3 50
—— Metalliferous Wastes..8vo,   *2 50
—— Analysis of Ashes and Alloys................................8vo,   *2 50
Parry, L. A. Risk and Dangers of Various Occupations.........8vo,   *4 25
Parshall, H. F., and Hobart, H. M. Armature Windings...........4to, *7 50
—— Electric Railway Engineering................................4to, *10 00
Parsons, J. L. Land Drainage..................................8vo,   *1 50
Parsons, S. J Malleable Cast Iron..............................8vo,   *2 50
Partington, J. R. Higher Mathematics for Chemical Students..12mo,   *2 00
—— Textbook of Thermodynamics................................8vo,   *4 00
Passmore, A. C. Technical Terms Used in Architecture.........8vo,   *4 25
Patchell, W. H. Electric Power in Mines.......................8vo,   *4 00
Paterson, G. W. L. Wiring Calculations.......................12mo,   *3 00
—— Electric Mine Signalling Installations.....................12mo,   *1 50
Patterson, D. The Color Printing of Carpet Yarns..............8vo,   *4 25
—— Color Matching on Textiles..................................8vo,   *4 25
—— Textile Color Mixing..8vo,   *4 25
Paulding, C. P. Condensation of Steam in Covered and Bare Pipes .8vo, *2 00
—— Transmission of Heat through Cold-storage Insulation.......12mo,   *1 00
Payne, D. W. Iron Founders' Handbook........................8vo,   *4 00
Peckham, S. F. Solid Bitumens................................8vo,   *5 00
Peddie, R. A. Engineering and Metallurgical Books............12mo,   *1 50
Peirce, B. System of Analytic Mechanics.......................4to, 10 00
—— Linnear Associative Algebra.................................4to,   3 00
Pendred, V. The Railway Locomotive. (Westminster Series.).....8vo, *2 00

Perkin, F. M. Practical Methods of Inorganic Chemistry........12mo,   *1 00
Perrin, J. Atoms..8vo,   *2 50
——— and Jaggers, E. M. Elementary Chemistry................12mo,   *1 00
Perrine, F. A. C. Conductors for Electrical Distribution...........8vo,   *3 50
Petit, G. White Lead and Zinc White Paints.................8vo,   *2 50
Petit, R. How to Build an Aeroplane. Trans. by T. O'B. Hubbard, and
    J. H. Ledeboer 8vo,   *1 50
Pettit, Lieut. J. S. Graphic Processes. (Science Series No. 76.)...16mo,   0 50
Philbrick, P. H. Beams and Girders. (Science Series No. 88.)...16mo,
Phillips, J. Gold Assaying.....................................8vo,   *3 75
——— Dangerous Goods...8vo,   3 50
Phin, J. Seven Follies of Science........................ ....12mo,   *1 25
Pickworth, C. N. The Indicator Handbook. Two Volumes..12mo, each,   1 50
——— Logarithms for Beginners...................... 12mo. boards,   0 50
——— The Slide Rule..12mo,   1 25
Pilcher, R. B., and Butler-Jones, F. What Industry Owes to Chemical
    Science........ ....12mo,   1 50
Plattner's Manual of Blow-pipe Analysis. Eighth Edition, revised. Trans.
    by H. B. Cornwall...8vo,   *4 00
Plympton, G. W. The Aneroid Barometer. (Science Series No. 35.) 16mo,   0 50
——— How to become an Engineer. (Science Series No. 100.)......16mo,   0 50
——— Van Nostrand's Table Book. (Science Series No. 104.).......16mo,   0 50
Pochet, M. L. Steam Injectors. Translated from the French. (Science
    Series No. 29.)...16mo,   0 50
Pocket Logarithms to Four Places. (Science Series No. 65.).. ...16mo,   0 50
                                                            leather,   1 00
Polleyn, F. Dressings and Finishings for Textile Fabrics...........8vo,   *3 00
Pope, F. G. Organic Chemistry...............................12mo,   2 50
Pope, F. L. Modern Practice of the Electric Telegraph...........8vo,   1 50
Popplewell, W. C. Prevention of Smoke........................8vo,   *4 25
——— Strength of Materials.....................................8vo,   *2 50
Porritt, B. D. The Chemistry of Rubber. (Chemical Monographs,
    No. 3.)..12mo,   *1 00
Porter, J. R. Helicopter Flying Machine................ .. ....12mo,   1 50
Potts, H. E. Chemistry of the Rubber Industry. (Outlines of Indus-
    trial Chemistry)...8vo,   *2 50
Practical Compounding cf Oils, Tallow and Grease............8vo,   *4 25
Pratt, K. Boiler Draught.. ....12mo,   *1 25
——— High Speed Steam Engines................................8vo,   *2 00
Pray, T., Jr. Twenty Years with the Indicator...................8vo,   2 50
——— Steam Tables and Engine Constant.........................8vo,   2 00
Prelini, C. Earth and Rock Excavation.........................8vo,   *3 00
——— Graphical Determination of Earth Slopes................ .....8vo,   *2 00
——— Tunneling. New Edition..................................8vo,   *3 00
——— Dredging. A Practical Treatise...........................8vo,   *3 00
Prescott, A. B. Organic Analysis..............................8vo,   5 00
Prescott, A. B., and Johnson, O. C. Qualitative Chemical Analysis. 8vo,   *3 50
Prescott, A. B., and Sullivan, E. C. First Book in Qualitative Chemistry.
                                                            12mo,   *1 50
Prideaux, E B. R. Problems in Physical Chemistry.............8vo,   *2 00
——— The Theory and Use of Indicators.........................8vo,   5 00
Primrose, G. S. C. Zinc. (Metallurgy Series.).........(In Press.)

Prince, G. T. Flow of Water..............................12mo,   *2 00
Pullen, W. W. F. Application of Graphic Methods to the Design of
    Structures..12mo,   *2 50
——Injectors: Theory, Construction and Working............12mo,   *2 00
——Indicator Diagrams 8vo,   *2 50
—— Engine Testing 8vo,   *5 50
Putsch, A. Gas and Coal-dust Firing..........................8vo,   *3 00
Pynchon, T. R. Introduction to Chemical Physics................8vo,   3 00

Rafter G. W. Mechanics of Ventilation. (Science Series No. 33.).16mo,   0 50
—— Potable Water. (Science Series No. 103.)................16mo,   0 50
—— Treatment of Septic Sewage. (Science Series No. 118.)...16mo,   0 50
Rafter, G. W., and Baker, M. N. Sewage Disposal in the United States.
                                    4to,   *6 00
Raikes, H. P. Sewage Disposal Works.........................8vo,   *4 00
Randau, P. Enamels and Enamelling..........................8vo,   *7 25
Rankine, W. J. M. Applied Mechanics........................8vo,   5 00
—— Civil Engineering..8vo,   6 50
—— Machinery and Millwork...................................8vo,   5 00
— — The Steam-engine and Other Prime Movers.................8vo,   5 00
Rankine, W. J. M., and Bamber, E. F. A Mechanical Text-book....8vo,   3 50
Ransome, W. R. Freshman Mathematics......................12mo,   *1 35
Raphael, F. C. Localization of Faults in Electric Light and Power Mains.
                                    8vo,   3 50
Rasch, E. Electric Arc Phenomena. Trans. by K. Tornberg.......8vo,   *2 00
Rathbone, R. L. B. Simple Jewellery..........................8vo,   *2 00
Rateau, A. Flow of Steam through Nozzles and Orifices. Trans. by H.
    B. Brydon..8vo   *1 50
Rausenberger, F. The Theory of the Recoil Guns...............8vo,   *5 00
Rautenstrauch, W. Notes on the Elements of Machine Design.8vo, boards,   *1 50
Rautenstrauch, W., and Williams, J. T. Machine Drafting and Empirical
    Design.
        Part I. Machine Drafting.................................8vo,   *1 25
        Part II. Empirical Design.......................(In Preparation.)
Raymond, E. B. Alternating Current Engineering.............12mo,   *2 50
Rayner, H. Silk Throwing and Waste Silk Spinning..............8vo,
Recipes for the Color, Paint, Varnish, Oil, Soap and Drysaltery Trades,
                                    8vo,   *6 50
Recipes for Flint Glass Making...............................12mo,   *5 25
Redfern, J. B., and Savin, J. Bells, Telephones (Installation Manuals
    Series.)...16mo,   *0 50
Redgrove, H. S. Experimental Mensuration..................12mo,   *1 25
Redwood, B. Petroleum. (Science Series No. 92.)..............16mo,   0 50
Reed, S. Turbines Applied to Marine Propulsion.................   *5 00
Reed's Engineers' Handbook.... ....8vo,   *9 00
—— Key to the Nineteenth Edition of Reed's Engineers' Handbook..8vo,   4 00
—— Useful Hints to Sea-going Engineers....................12mo,   3 00
Reid, E. E. Introduction to Research in Organic Chemistry. (In Press.)
Reid, H. A. Concrete and Reinforced Concrete Construction......8vo,   *5 00
Reinhardt, C. W. Lettering for Draftsmen, Engineers, and Students.
                    oblong 4to, boards,   1 00

Reinhardt, C. W. The Technic of Mechanical Drafting,
oblong, 4to, boards,   *1 00
Reiser, F. Hardening and Tempering of Steel. Trans. by A. Morris and
H. Robson 12mo,   *3 00
Reiser, N. Faults in the Manufacture of Woolen Goods. Trans. by A.
Morris and H. Robson................................8vo,   *3 00
——— Spinning and Weaving Calculations.......................8vo,   *6 25
Renwick, W. G. Marble and Marble Working...................8vo,   5 00
Reuleaux, F. The Constructor. Trans. by H. H. Suplee.........4to,   *4 00
Reuterdahl, A. Theory and Design of Reinforced Concrete Arches.8vo,   *2 00
Rey, Jean. The Range of Electric Searchlight Projectors........8vo,   *4 50
Reynolds, O., and Idell, F. E. Triple Expansion Engines. (Science
Series No. 99.) 16mo,   0 50
Rhead, G. F. Simple Structural Woodwork...................12mo,   *1 25
Rhodes, H. J. Art of Lithography............................8vo,   6 50
Rice, J. M., and Johnson, W. W. A New Method of Obtaining the Differ-
ential of Functions...............................12mo,   0 50
Richards, W. A. Forging of Iron and Steel...................12mo,   1 50
Richards, W. A., and North, H. B. Manual of Cement Testing....12mo,   *1 50
Richardson, J. The Modern Steam Engine.....................8vo,   *3 50
Richardson, S. S. Magnetism and Electricity...................12mo,   *2 00
Rideal, S. Glue and Glue Testing............................8vo,   *6 50
Riesenberg, F. The Men on Deck...........................12mo,   3 00
Rimmer, E. J. Boiler Explosions, Collapses and Mishaps..........8vo,   *1 75
Rings, F. Concrete in Theory and Practice...................12mo,   *2 50
——— Reinforced Concrete Bridges............................4to,   *5 00
Ripper, W. Course of Instruction in Machine Drawing...........folio,   *6 00
Roberts, F. C. Figure of the Earth. (Science Series No. 79.).....16mo,   0 50
Roberts, J., Jr. Laboratory Work in Electrical Engineering.......8vo,   *2 00
Robertson, L. S. Water-tube Boilers.........................8vo,   2 00
Robinson, J. B. Architectural Composition.....................8vo,   *2 50
Robinson, S. W. Practical Treatise on the Teeth of Wheels. (Science
Series No. 24.)......................................16mo,   0 50
—— Railroad Economics. (Science Series No. 59.)............16mo,   0 50
——— Wrought Iron Bridge Members. (Science Series No. 60.).....16mo,   0 50
Robson, J. H. Machine Drawing and Sketching.................8vo,   *2 00
Roebling, J. A. Long and Short Span Railway Bridges.........folio,   25 00
Rogers, A. A Laboratory Guide of Industrial Chemistry........8vo,   *2 00
——— Elements of Industrial Chemistry........................12mo,   *3 00
——— Manual of Industrial Chemistry.........................8vo,   *5 00
Rogers, F. Magnetism of Iron Vessels. (Science Series No. 30.).16mo,   0 50
Rohland, P. Colloidal and Crystalloidal State of Matter. Trans. by
W. J. Britland and H. E. Potts........................12mo,   *1 25
Rollinson, C. Alphabets...........................Oblong, 12mo,   *1 00
Rose, J. The Pattern-makers' Assistant........................8vo,   2 50
—— Key to Engines and Engine-running......................12mo,   2 50
Rose, T. K. The Precious Metals. (Westminster Series.) 8vo,   *2 00
Rosenhain, W. Glass Manufacture. (Westminster Series.)......8vo,   *2 00
——— Physical Metallurgy, An Introduction to. (Metallurgy Series.)
8vo,   *3 50
Roth, W. A. Physical Chemistry............................8vo,   *2 00

Rowan, F. J. Practical Physics of the Modern Steam-boiler......8vo, *3 00
—— and Idell, F. E. Boiler Incrustation and Corrosion. (Science
        Series No. 27.)..16mo,   o 50
Roxburgh, W. General Foundry Practice. (Westminster Series.).8vo, *2 00
Ruhmer, E. Wireless Telephony. Trans. by J. Erskine-Murray..8vo, *4 50
Russell, A. Theory of Electric Cables and Networks.............8vo, *3 00
Rutley, F. Elements of Mineralogy...........................12mo, *1 25

Sandeman, E. A. Notes on the Manufacture of Earthenware...12mo,   3 50
Sanford, P. G. Nitro-explosives..............................8vo, *4 00
Saunders, C. H. Handbook of Practical Mechanics............16mo,   1 00
                                                            leather,   1 25
Sayers, H. M. Brakes for Tram Cars.........................8vo, *1 25
Scheele, C. W. Chemical Essays.............................8vo, *2 00
Scheithauer, W. Shale Oils and Tars........................8vo, *5 00
Scherer, R. Casein. Trans. by C. Salter....................8vo, *4 25
Schidrowitz, P. Rubber, Its Production and Industrial Uses......8vo, *6 00
Schindler, K. Iron and Steel Construction Works.............12mo, *2 25
Schmall, C. N. First Course in Analytic Geometry, Plane and Solid.
                                        12mo, half leather, *1 75
Schmeer, L. Flow of Water.................................8vo, *3 00
Schumann, F. A Manual of Heating and Ventilation....12mo, leather,   1 50
Schwarz, E. H. L. Causal Geology.........................8vo, *3 00
Schweizer, V. Distillation of Resins........................8vo,   4 50
Scott, W. W. Qualitative Analysis. A Laboratory Manual. New
        Edition...   2 50
—— Standard Methods of Chemical Analysis.................8vo, *6 00
Scribner, J. M. Engineers' and Mechanics' Companion..16mo, leather,   1 50
Scudder, H. Electrical Conductivity and Ionization Constants of
        Organic Compounds.................................8vo, *3 00
Searle, A. B. Modern Brickmaking.........................8vo, *7 25
—— Cement, Concrete and Bricks..........................8vo, *6 50
Searle, G. M. "Sumners' Method." Condensed and Improved.
        (Science Series No. 124.).........................16mo,   o 50
Seaton, A. E. Manual of Marine Engineering.................8vo   8 00
Seaton, A. E., and Rounthwaite, H. M. Pocket-book of Marine Engi-
        neering....... ....16mo, leather,   3 50
Seeligmann, T., Torrilhon, G. L., and Falconnet, H. India Rubber and
        Gutta Percha. Trans. by J. G. McIntosh..............8vo, *5 00
Seidell, A. Solubilities of Inorganic and Organic Substances....8vo,   3 00
Seligman, R. Aluminum. (Metallurgy Series.)........(In Press.)
Sellew, W. H. Steel Rails..................................4to, *10 00
—— Railway Maintenance Engineering.......................12mo, *2 50
Senter, G. Outlines of Physical Chemistry..................12mo, *2 00
—— Text-book of Inorganic Chemistry.......................12mo, *2 00
Sever, G. F. Electric Engineering Experiments..........8vo, boards, *1 00
Sever, G. F., and Townsend, F. Laboratory and Factory Tests in Elec-
        trical Engineering....... ....8vo, *2 50
Sewall, C. H. Wireless Telegraphy.........................8vo, *2 00
—— Lessons in Telegraphy................................12mo, *1 00

Sewell, T. The Construction of Dynamos......................8vo, *3 00
Sexton, A. H. Fuel and Refractory Materials.............12mo, *2 50
——— Chemistry of the Materials of Engineering................12mo, *2 50
——— Alloys (Non-Ferrous).......................................8vo, *3 00
Sexton, A. H., and Primrose, J. S. G. The Metallurgy of Iron and Steel.
                                                          8vo, *6 50
Seymour, A. Modern Printing Inks...........................8vo, *3 00
Shaw, Henry S. H. Mechanical Integrators. (Science Series No. 83.)
                                                          16mo, 0 50
Shaw, S. History of the Staffordshire Potteries.................8vo, 3 00
——— Chemistry of Compounds Used in Porcelain Manufacture....8vo, *6 00
Shaw, T. R. Driving of Machine Tools......................12mo, *2 50
——— Precision Grinding Machines..........................12mo, 5 50
Shaw, W. N. Forecasting Weather...........................8vo, *3 50
Sheldon, S., and Hausmann, E. Direct Current Machines........12mo, *2 50
——— Alternating Current Machines.........................12mo, *2 50
Sheldon, S., and Hausmann, E. Electric Traction and Transmission
        Engineering.......................................12mo, *2 50
——— Physical Laboratory Experiments, for Engineering Students..8vo, *1 25
Shields, J. E. Notes on Engineering Construction.............12mo, 1 50
Shreve, S. H. Strength of Bridges and Roofs...................8vo, 3 50
Shunk, W. F. The Field Engineer................12mo, fabrikoid, 2 50
Simmons, W. H., and Appleton, H. A. Handbook of Soap Manufacture,
                                                          8vo, *5 00
Simmons, W. H., and Mitchell, C. A. Edible Fats and Oils......8vo, *4 50
Simpson, G. The Naval Constructor.................12mo, fabrikoid, *5 00
Simpson, W. Foundations........................8vo. (In Press.)
Sinclair, A. Development of the Locomotive Engine...8vo, half leather, 5 00
Sindall, R. W. Manufacture of Paper. (Westminster Series.)....8vo, *2 00
Sindall, R. W., and Bacon, W. N. The Testing of Wood Pulp.....8vo, *2 50
Sloane, T. O'C. Elementary Electrical Calculations...........12mo, *2 00
Smallwood, J. C. Mechanical Laboratory Methods. (Van Nostrand's
        Textbooks.) ....12mo, fabrikoid, *3 00
Smith, C. A. M. Handbook of Testing, MATERIALS............8vo, *2 50
Smith, C. A. M., and Warren, A. G. New Steam Tables.........8vo, *1 25
Smith, C. F. Practical Alternating Currents and Testing........8vo, *3 50
——— Practical Testing of Dynamos and Motors...............8vo, *3 00
Smith, F. A. Railway Curves..............................12mo, *1 00
——— Standard Turnouts on American Railroads.............12mo, *1 00
——— Maintenance of Way Standards.......................12mo, *1 50
Smith, F. E. Handbook of General Instruction for Mechanics...12mo, 1 50
Smith, G. C. Trinitrotoluenes and Mono- and Dinitrotoluenes, Their
        Manufacture and Properties.......................12mo, 2 00
Smith, H. G. Minerals and the Microscope.....................12mo, *1 25
Smith, J. C. Manufacture of Paint...........................8vo, *3 50
Smith, R. H. Principles of Machine Work....................12mo,
——— Advanced Machine Work..............................12mo, *3 00
Smith, W. Chemistry of Hat Manufacturing.................12mo, *4 50
Snell, A. T. Electric Motive Power..........................8vo, *4 00
Snow, W. G. Pocketbook of Steam Heating and Ventilation. (In Press.)
Snow, W. G., and Nolan, T. Ventilation of Buildings. (Science Series
        No. 5.)..16mo, 0 50
Soddy, F. Radioactivity....................................8vo, *3 00

Solomon, M. Electric Lamps. (Westminster Series.)...........8vo,
Somerscales, A. N. Mechanics for Marine Engineers...........12mo,
—— Mechanical and Marine Engineering Science.................8vo,
Sothern, J. W. The Marine Steam Turbine.....................8vo,
—— Verbal Notes and Sketches for Marine Engineers...........8vo,
Sothern, J. W., and Sothern, R. M. Elementary Mathematics for
    Marine Engineers.......................................12mo,
—— Simple Problems in Marine Engineering Design............12mo,
Southcombe, J. E. Chemistry of the Oil Industries   (Outlines of In-
    dustrial Chemistry.)..................................8vo,
Soxhlet, D. H. Dyeing and Staining Marble. Trans. by A. Morris and
    H. Robson 8vo,
Spangenburg, L. Fatigue of Metals. Translated by S. H. Shreve.
    (Science Series No. 23.)..............................16mo,
Specht, G. J., Hardy, A. S., McMaster, J. B., and Walling. Topographical
    Surveying. (Science Series No. 72.).................16mo,
Spencer, A. S. Design of Steel-Framed Sheds.................8vo,
Speyers, C. L. Text-book of Physical Chemistry.............8vo,
Spiegel, L. Chemical Constitution and Physiological Action. (Trans.
    by C. Luedeking and A. C. Boylston.)..................12mo,
Sprague, E. H. Hydraulics....................................12mo,
—— Elements of Graphic Statics...............................8vo,
—— Stability of Masonry......................................12mo,
—— Elementary Mathematics for Engineers.....................12mo,
—— Stability of Arches.......................................12mo,
—— Strength of Structural Elements...........................12mo,
Stahl, A. W. Transmission of Power. (Science Series No. 28.) .16mo,
Stahl, A. W., and Woods, A. T. Elementary Mechanism 12mo,
Staley, C., and Pierson, G. S. The Separate System of Sewerage...8vo,
Standage, H. C. Leatherworkers' Manual......................8vo,
—— Sealing Waxes, Wafers, and Other Adhesives..............8vo,
—— Agglutinants of all Kinds for all Purposes.................12mo,
Stanley, H. Practical Applied Physics..................(In Press.)
Stansbie, J. H. Iron and Steel. (Westminster Series.)..........8vo,
Steadman, F. M. Unit Photography...........................12mo,
Stecher, G. E. Cork. Its Origin and Industrial Uses.........12mo,
Steinman, D. B. Suspension Bridges and Cantilevers. (Science Series
    No. 127.)...
—— Melan's Steel Arches and Suspension Bridges.................8vo,
Stevens, E. J. Field Telephones and Telegraphs.................
Stevens, H. P. Paper Mill Chemist 16mo,
Stevens, J. S. Theory of Measurements......................12mo,
Stevenson, J. L. Blast-Furnace Calculations..........12mo, leather,
Stewart, G. Modern Steam Traps.............................12mo,
Stiles, A. Tables for Field Engineers.......................12mo,
Stodola, A. Steam Turbines. Trans. by L. C. Loewenstein.......8vo,
Stone, H. The Timbers of Commerce...........................8vo,
Stopes, M. Ancient Plants...................................8vo,
—— The Study of Plant Life...................................8vo,
Sudborough, J. J., and James, T. C. Practical Organic Chemistry..12mo,
Suffling, E. R. Treatise on the Art of Glass Painting.........8vo,
Sullivan, T. V., and Underwood, N. Testing and Valuation of Build-
    ing and Engineering Materials...................(In Press.)

Sur, F. J. S. Oil Prospecting and Extracting...................8vo,   *1 00
Svenson, C. L. Handbook on Piping...........................8vo,   4 00
—— Essentials of Drafting.....................................8vo,   1 50
Swan, K. Patents, Designs and Trade Marks. (Westminster Series.).
                                                            8vo,   *2 00
Swinburne, J., Wordingham, C. H., and Martin, T. C. Electric Currents.
      (Science Series No. 109.)............................16mo,   0 50
Swoope, C. W. Lessons in Practical Electricity...............12mo,   *2 00

Tailfer, L. Bleaching Linen and Cotton Yarn and Fabrics......8vo,   8 50
Tate, J. S. Surcharged and Different Forms of Retaining-walls. (Science
      Series No. 7.)..16mo,   0 50
Taylor, F. N. Small Water Supplies..........................12mo,   *2 50
—— Masonry in Civil Engineering.............................8vo,   *2 50
Taylor, T. U. Surveyor's Handbook...............12mo, leather,   *2 00
—— Backbone of Perspective..................................12mo,   *1 00
Taylor, W. P. Practical Cement Testing......................8vo,   *3 00
Templeton, W. Practical Mechanic's Workshop Companion.
                                            12mo, morocco,   2 00
Tenney, E. H. Test Methods for Steam Power Plants. (Van
      Nostrand's Textbooks.) 12mo,   *2 50
Terry, H. L. India Rubber and its Manufacture. (Westminster Series.)
                                                            8vo,   *2 00
Thayer, H. R. Structural Design. 8vo.
      Vol. I. Elements of Structural Design.......................   *2 00
      Vol. II. Design of Simple Structures........................   *4 00
      Vol. III. Design of Advanced Structures........(In Preparation.)
—— Foundations and Masonry...................(In Preparation.)
Thiess, J. B., and Joy, G. A. Toll Telephone Practice...........8vo,   *3 50
Thom, C., and Jones, W. H. Telegraphic Connections....oblong, 12mo,   1 50
Thomas, C. W. Paper-makers' Handbook................(In Press.)
Thomas, J. B. Strength of Ships.............................8vo,   3 00
Thomas, Robt. G. Applied Calculus..............12mo (In Press.)
Thompson, A. B. Oil Fields of Russia 4to,   *7 50
—— Oil Field Development....................................   7 50
Thompson, S. P. Dynamo Electric Machines. (Science Series No. 75.)
                                                            16mo,   0 50
Thompson, W. P. Handbook of Patent Law of All Countries.....16mo,   1 50
Thomson, G. Modern Sanitary Engineering...................12mo,   *3 00
Thomson, G. S. Milk and Cream Testing....................12mo,   *2 25
—— Modern Sanitary Engineering, House Drainage, etc..........8vo,   *3 00
Thornley, T. Cotton Combing Machines......................8vo,   *3 00
—— Cotton Waste...8vo,   *4 50
—— Cotton Spinning. 8vo.
      First Year *2 00
      Second Year *4 25
      Third Year *3 50
Thurso, J. W. Modern Turbine Practice.......................8vo,   *4 00
Tidy, C. Meymott. Treatment of Sewage. (Science Series No. 94.)16mo,   0 50
Tillmans, J. Water Purification and Sewage Disposal. Trans. by
      Hugh S. Taylor..8vo,   *2 00
Tinney, W. H. Gold-mining Machinery.......................8vo,   *3 00
Titherley, A. W. Laboratory Course of Organic Chemistry.......8vo,   *2 00

Tizard, H. T. Indicators..................................(*In Press.*)
Toch, M. Chemistry and Technology of Paints................8vo,   *4 00
—— Materials for Permanent Painting.......................12mo,   *2 00
Tod, J., and McGibbon, W. C. Marine Engineers' Board of Trade
    Examinations 8vo,   *2 00
Todd, J., and Whall, W. B. Practical Seamanship..............8vo,   8 00
Tonge, J. Coal. (Westminster Series.).......................8vo,   *2 00
Townsend, F. Alternating Current Engineering..........8vo, boards,   *0 75
Townsend, J. S. Ionization of Gases by Collision..............8vo,   *1 25
Transactions of the American Institute of Chemical Engineers, 8vo.
    Eight volumes now ready. Vol. I. to IX., 1908-1916. Vol.
    X. *In Press*...8vo, each,   6 00
Traverse Tables. (Science Series No. 115.)..................16mo,   0 50
                       morocco,   1 00
Treiber, E. Foundry Machinery. Trans. by C. Salter........ 12mo,   1 50
Trinks, W., and Housum, C. Shaft Governors. (Science Series No. 122.)
                        16mo,   0 50
Trowbridge, W. P. Turbine Wheels. (Science Series No. 44.)..16mo,   0 50
Tucker, J. H. A Manual of Sugar Analysis....................8vo,   3 50
Tunner, P. A. Treatise on Roll-turning. Trans. by J. B. Pearse.
                  8vo, text and folio atlas,   10 00
Turnbull, Jr., J., and Robinson, S. W. A Treatise on the Compound
    Steam-engine. (Science Series No. 8.)...............16mo,
Turner, H. Worsted Spinners' Handbook......................12mo,   *3 50
Turrill, S. M. Elementary Course in Perspective............ 12mo,   *1 25
Twyford, H. B. Purchasing...................................8vo,   *3 00
—— Storing, Its Economic Aspects and Proper Methods..........8vo,   3 50
Tyrrell, H. G. Design and Construction of Mill Buildings.......8vo, *4 00
—— Concrete Bridges and Culverts...................16mo, leather,   *3 00
—— Artistic Bridge Design..................................:.8vo,   *3 00

Underhill, C. R. Solenoids, Electromagnets and Electromagnetic Wind-
    ings...12mo,   *2 00
Underwood, N., and Sullivan, T. V. Chemistry and Technology of
    Printing Inks 8vo,   *3 00
Urquhart, J. W. Electro-plating............................12mo,   2 00
—— Electrotyping..12mo,   2 00
Usborne, P. O. G. Design of Simple Steel Bridges.............8vo,   *4 00

Vacher, F. Food Inspector's Handbook 12mo,
Van Nostrand's Chemical Annual. Fourth issue 1918.fabrikoid, 12mo, *3 00
—— Year Book of Mechanical Engineering Data.........(*In Press.*)
Van Wagenen, T. F. Manual of Hydraulic Mining.............16mo,   1 00
Vega, Baron Von. Logarithmic Tables.......................8vo,   2 50
Vincent, C. Ammonia and its Compounds. Trans. by M. J. Salter.8vo, *3 00
Volk, C. Haulage and Winding Appliances....................8vo,   *4 00
Von Georgievics, G. Chemical Technology of Textile Fibres. Trans.
    by C. Salter..8vo,
—— Chemistry of Dyestuffs. Trans. by C. Salter..............8vo,   *4 50
Vose, G. L. Graphic Method for Solving Certain Questions in Arithmetic
    and Algebra (Science Series No. 16.)...............16mo,   0 50

Vosmaer, A. Ozone...8vo,   *2 50

Wabner, R. Ventilation in Mines. Trans. by C. Salter.........8vo,   *6 50
Wade, E. J. Secondary Batteries.............................8vo,   *4 00
Wadmore, T. M. Elementary Chemical Theory...............12mo,   *1 50
Wagner, E. Preserving Fruits, Vegetables, and Meat.........12mo,   *3 00
Wagner, J. B. A Treatise on the Natural and Artificial Processes of
    Wood Seasoning.......................................8vo,   3 00
Waldram, P. J. Principles of Structural Mechanics............12mo,   *3 00
Walker, F. Dynamo Building. (Science Series No. 98.)........16mo,   0 50
Walker, J. Organic Chemistry for Students of Medicine........8vo,   *3 00
Walker, S. F. Steam Boilers, Engines and Turbines...........8vo,   3 00
—— Refrigeration, Heating and Ventilation on Shipboard.......12mo,   *2 00
—— Electricity in Mining.....................................8vo,   *4 50
Wallis-Tayler, A. J. Bearings and Lubrication.................8vo,   *1 50
—— Aerial or Wire Ropeways...................................8vo,   *3 00
—— Preservation of Wood......................................8vo,   4 00
—— Refrigeration, Cold Storage and Ice Making...............8vo,   5 50
—— Sugar Machinery...12mo,   *2 50
Walsh, J. J. Chemistry and Physics of Mining and Mine Ventilation,
                                                           12mo,   *2 00
Wanklyn, J. A. Water Analysis...............................12mo,   2 00
Wansbrough, W. D. The A B C of the Differential Calculus....12mo,   *2 50
—— Slide Valves...12mo,   *2 00
Waring, Jr., G. E. Sanitary Conditions. (Science Series No. 31.).16mo,   0 50
—— Sewerage and Land Drainage................................   *6 00
—— Modern Methods of Sewage Disposal.....................12mo,   2 00
—— How to Drain a House...................................12mo,   1 25
Warnes, A. R. Coal Tar Distillation...........................8vo,   *5 00
Warren, F. D. Handbook on Reinforced Concrete...............12mo,   *2 50
Watkins, A. Photography. (Westminster Series.)..............8vo,   *2 00
Watson, E. P. Small Engines and Boilers.....................12mo,   1 25
Watt, A. Electro-plating and Electro-refining of Metals........8vo,   *4 50
—— Electro-metallurgy.......................................12mo,   1 00
—— The Art of Soap Making..................................8vo,   3 00
—— Leather Manufacture.......................................8vo,   *4 00
—— Paper-Making..8vo,   3 00
Webb, H. L. Guide to the Testing of Insulated Wires and Cables.12mo,   1 00
Webber, W. H. Y. Town Gas. (Westminster Series.)..........8vo,   *2 00
Wegmann, Edward. Conveyance and Distribution of Water for
    Water Supply...8vo,   5 00
Weisbach, J. A Manual of Theoretical Mechanics..............8vo,   *6 00
                                                           sheep,   *7 50
Weisbach, J., and Herrmann, G. Mechanics of Air Machinery....8vo,   *3 75
Wells, M. B. Steel Bridge Designing.........................8vo,   *2 50
Wells, Robt. Ornamental Confectionery.....................12mo,   3 00
Weston, E. B. Loss of Head Due to Friction of Water in Pipes..12mo,   *1 50
Wheatley, O. Ornamental Cement Work........................8vo,   *2 25
Whipple, S. An Elementary and Practical Treatise on Bridge Building.
                                                            8vo,   3 00
White, C. H. Methods of Metallurgical Analysis. (Van Nostrand's
    Textbooks.) 12mo,   2 50

www.ingramcontent.com/pod-product-compliance
Lightning Source LLC
LaVergne TN
LVHW012203040326
832903LV00003B/96